FINE MARBLE IN ARCHITECTURE

FINE MARBLE IN ARCHITECTURE

STUDIO MARMO

TEXT BY FREDERICK BRADLEY

W. W. NORTON & COMPANY, INC.
NEW YORK • LONDON

Copyright © 2001 by Absolute Printers
Copyright © 2001 by W.W. Norton & Company, Inc.

First American edition published by W.W. Norton & Company, Inc., 2001

First published in Italy by Absolute Printers 2000 under the title MARMI PREGIATI IN ARCHITETTURA/ FINE MARBLES IN ARCHITECTURE

All rights reserved
Printed in Italy

For information about permission to reproduce selections from this book, write to Permissions, W.W. Norton & Company, Inc., 500 Fifth Avenue, New York, NY 10110

The text of this book is composed in Gill Sans
with the display set in Gill Sans Bold and Gill Sans Bold Condensed
Composition by Gina Webster
Manufacturing by Pacini Editore S.p.A. (Pisa)

Library of Congress Cataloging-in-Publication Data
Bradley, Frederick.
Fine marble in architecture / Studio Marmo ; text by Frederick Bradley.
p. cm.
Published also in Italian as Marmi pregiati in architettura.
Includes bibliographical references and index.
ISBN 0-393-73074-3
1. Building, Stone. 2. Building stones. 3. Marble. 4. Architecture. I. Studio Marmo (Florence, Italy) II. Title
TH1201 .B73 2001
721'.0441–dc21
00-069944
CIP

W.W. Norton & Company, Inc., 500 Fifth Avenue, New York, NY 10110
www.wwnorton.com

W.W. Norton & Company Ltd., 10 Coptic Street, London WC1A 1PU

0 9 8 7 6 5 4 3 2 1

CONTENTS

Introduction   vii

PART I • MARBLE: THE BACKGROUND   9

   Chapter 1 • Marble: The Raw Material   11
   Chapter 2 • Marble: The Premium Material   19
   Chapter 3 • Working and Using Premium Marble   29
   Chapter 4 • Physical-Mechanical Data   39

PART II • THE MARBLES   49

PART III • PORTFOLIO OF BUILT WORKS   147

Bibliography   190

Index of Fine Marbles   191

About the CD-ROM   192

INTRODUCTION

The use of stone in architecture for decorative purposes is a widespread phenomenon. The need to cover large surfaces (often fairly inaccessible to direct observation) has increasingly led to the use of materials which, in addition to being available in large quantities, show good constancy in their esthetic features, without necessarily being of high quality. And so ornamental stone materials passed from their original role as a decorative element enriching an architectural work to a functional one, responding primarily to technical rather than purely esthetic demands.

At the start of the third millennium, however, preponderant interest seems to be in varieties of ornamental stone whose characteristics differ from those for standardized, industrial-type use and are, on the contrary, prevalently high in quality, especially in terms of esthetic worth.

This book illustrates this clear countertrend in the ornamental stone market and is meant to be a spokesman for it and, even more, to directly contribute to its spread.

PART I

MARBLE: THE BACKGROUND

CHAPTER 1
MARBLE: THE RAW MATERIAL

TERMINOLOGY AND CLASSIFICATION

The ancient Romans divided stone materials for building purposes into two big groups: the materials that could be polished called *marmora* (from the Greek *marmarios*, which means shining) and nonpolishable stones called *lapis*.

While this classification is rather broad, it continues to have a raison d'être in terms of stone workability. Even nowadays the market generically views **stones** (the Roman *lapis*) as many materials of very different origin and composition whose common denominator (except in some rare cases) is that they cannot be polished.

The discourse is obviously different for polishable materials. On the basis of geological knowledge, what was once generically considered *marmora* is now divided into at least three distinct commercial groups: **marbles, granites,** and **quartzites,** which refer to materials that in most cases have certain specific chemical, mineralogical, and physical-mechanical characteristics.

Without going into detail, we will list here the guidelines of this division and describe the salient features of each group.

Marbles primarily include materials of a carbonate nature, essentially calcium carbonate but also, to a lesser degree, a mixture of calcium carbonate and magnesium. Among these are marbles in the petrographical meaning of the term, that is, carbonate lithotypes of a meta-

Photo page 8: Door of the Sala del Consiglio, Palazzo Vecchio, Florence. Benedetto da Maino, circa 1470.
Photo facing page: Façade of Santa Maria Novella, Florence. Leon Battista Alberti, circa 1440.

morphic origin (such as *Bianco Carrara, Rosa Portugal, Imperial Danby*) and limestones of a sedimentary origin, either detrital (*Botticino, Breccia Pernice*) or chemical (*Travertino Romano, Green Onyx*). The commercial group of marbles also includes rocks of a silicate nature and magmatic origin like the serpentinites (*Serpentino Classico*) and the ophicalcites (*Rosso Levanto*). This seemingly anomalous inclusion is due to the fact that these materials, although of silicate composition, have characteristics that enable them to be worked with methods and technologies similar to those used on carbonate materials.

Granites are solely lithotypes of a silicate nature, whether of a prevalently intrusive magmatic origin and only partly of a lode, or of metamorphic origin. In the first case, chemical composition varies from acid (*Grigio Sardo, Rosa Porrinho*) to intermediate (*Baltic Brown*) to basic (*Nero Africa, Nero Zimbabwe*). In the second, chemical composition is prevalently acid (*Multicolor Red, Kashmir White*).

The **quartzite** group includes lithotypes of a metamorphic origin that are composed essentially of granules of quartz (*Azul Macaubas, Azul Imperial*).

USE PREREQUISITES

In order to be used for ornamental purposes, a stone material must possess certain prerequisites that respond to commercial and work requirements.

Commercial characteristics
A stone material's commercial potential basically depends on the following factors:
- esthetic appearance of the final product
- availability of the raw material
- in addition to the above is cost, a factor not weighable in absolute terms

Esthetic appearance
Given the material's ornamental function, its appearance is generally the most important factor. The esthetic worth of each material derives from the combination of three distinct elements, which form a whole that can be evaluated overall: these are color, pattern, and grain size.

Color often plays an important part in defining the esthetic picture of a material and consequently its commercial value as well. Table 1.1 gives a division of stone materials by market

FINE MARBLE IN ARCHITECTURE

Table 1.1: Main Dimension Stone Materials, Classed According to Market Characteristics

MARKET	MARBLE	HOMOGENEOUS GRANITE	ORIENTED GRANITE	VEINED GRANITE	QUARTZITE	COLOR CATEGORIES
High quality, medium/high cost materials characterized by colors which are easy to use; often supply does not satisfy demand; availability is generally constant	White, Pink, Black, Red, Green	Red, Black, White, Brown, Blue	Red, Pink, Green, Yellow	Red, Pink		*Classic*
High quality, medium/high cost materials with limited uses compared to the classics; demand may vary considerably; in many cases availability is not constant	Violet, Yellow, Brown	Green, Yellow,	Brown	Green, Violet, Yellow	Sky blue, Pink, Green	*Special*
Average quality, relatively low cost materials; availability is generally very good and fairly constant.	Beige, Grey	Pink, Grey	Grey			*Common*

category on the basis of their color. This division may change in time since color is susceptible to variations in taste and because there is an enormous range of colors on the market, continually updated by the discovery of new commercial varieties. Identifying the groups of materials belonging to the different market categories is a prelude to the concept of premium marble, which will be discussed in later chapters.

The *pattern* of a stone material depends on the arrangement of its constituent elements and/or the presence of veining and how this runs over the ground mass. The pattern is *homogeneous* when its components are evenly distributed and there is no veining, so that it is configured as a visually isotropic whole on tile scale. The pattern is *oriented* when (still without veining) the components show a clearly directional arrangement that gives the material a definite polarity. Finally, a pattern is called *veined* when there is more or less accentuated veining, whose distribution may be uneven or take a prevalent direction and give the material polarity again.

As it appears in the final product, the pattern of a stone material can be heavily influenced by the block's sawing direction and its orientation with respect to the run of the veining and the material's components. In detail, a squared block with veining or an obvious orientation of its components will, on three sides at right angles to one another, often have patterns completely different from each other (fig. 1). Using insider jargon, the surface parallel to the run plane of the veining is called *grain*, the surface at right angles to it but running in the same direction is called *easy way* and the surface at right angles to the grain and the easy way is called *hard way*.

Very generally, it can be said that, if sawed with the grain, a material with well-oriented veining will show wide and very undulating veining, often uneven but in any case tending to form nearly circular figures. If sawed the easy way, the veins will appear narrower, longer, nearly parallel to one another, and not very undulating. Finally, if sawed the hard way, the veining will be much thinner, shorter, and in an irregular and practically rectilinear run.

The third factor determining a material's appearance is its *grain size*, or the size of the individual crystals and other elements comprising the rock. Grain size's influence on esthetics can be direct (when, for example, the size of the crystals makes them stand out and become essential elements in the overall look of the material, as in breccias and many granites) but also indirect, working to various degrees on the intensity of the ground's color and on definition of the veining. Given equal color tone, a fine-grain material has a more intense and deeper color than a medium- or large-grain one. Likewise, the veining of a fine-grain material will be more clear-cut and defined than the veining on a large-grain one, where the passage between the color of the veining and that of the ground is much more gradual and shaded.

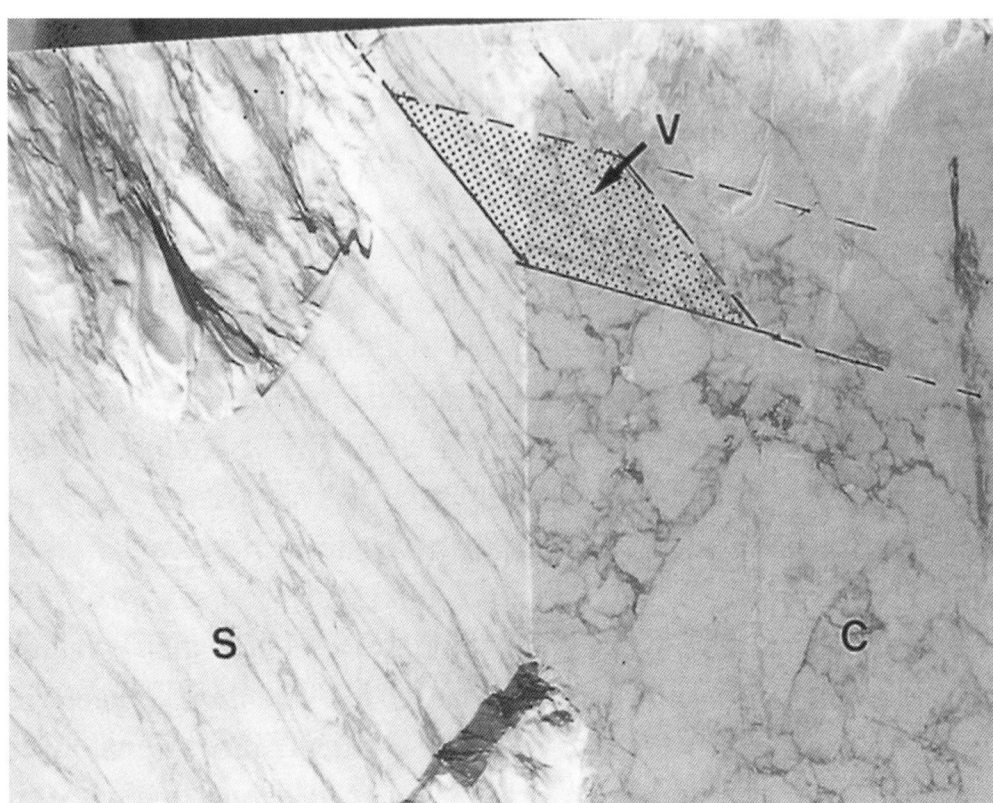

fig. 1:
v: grain;
s: easy way;
c: hard way

FINE MARBLE IN ARCHITECTURE

Availability

A material's availability indicates both the quantity that can be bought within a specific period of time and the shape and volume of the raw product. This latter parameter can vary heavily from case to case, so the market has different categories of raw product. By definition, the raw product is a *block squared* on 6 sides, devoid of cracks and sized to produce large slabs and optimize the space available beneath the gangsaw. Blocks of this type generally vary from 6 to 12 cubic meters. If some of the block's sides are not well-squared or are out of plumb or crossed by one or more cracks that could jeopardize slab yield, the block is called *semi-squared*. Highly fractured blocks, small and of no specific shape (and hence destined for tiles and small blocks) are known as shapeless blocks or, more simply, *shapeless*.

Workability characteristics

The workability characteristics of a stone material essentially depend on its physical-mechanical properties which, to various degrees, are tied to its mineralogical and chemical composition as well as to the rock's petrography. All these technical parameters can obviously be examined through specific laboratory tests, and their assessment is essential to understanding the real use possibilities for the material. Without going into the subject in detail, we refer the reader to Table 1.2 which shows the most common physical-mechanical tests and the weight they carry in normal uses of ornamental stone materials.

To conclude this brief introduction, we feel it is important to call attention to an aspect little dealt with in sector literature but, given this book's subject matter, a factor of importance. We refer to the defects stone materials may have and their possible influences on value. Apropos, it should be specified that, in practice, there are no commercial varieties devoid of defects and that the part of production without them—therefore considered of prime quality—is almost always just a small percentage of the total marketed. Commercial terminology commonly defines defects as both the characteristics that have a negative influence on the material's esthetics and the cracks in the blocks or worked products that can compromise their use. However, no univocal, officially recognized definition of defect exists, and there are varieties in which the borderline between a material's feature and its defect is very vague. This is the case with *Bianco Carrara*, in which the presence of tiny holes called *taroli* is so common that many consider this a feature (albeit a negative one) of the material and not a defect. The question, especially in light of the many legal diatribes arising from differences of opinion, is not one to underestimate.

Following a definition already given in the past, defects can be taken as all those physical

Table 1.2: Importance of the Physical-Mechanical Properties of Dimension Stone According to Its Use and Standard Laboratory Test Codes

	Weight per unit of volume	Compressive strength	Compressive strength after freezing	Imbibition	Flexural strength/ Modulus of rupture	Linear thermal expansion	Modulus of elasticity	Abrasion resistance	Impact strength	Knoop microhardness
Internal claddings	ooo	ooo	o	o	oo	o	o	o	o	o
External claddings	ooo	ooo	ooo	ooo	ooo	ooo	ooo	o	o	o
Internal floorings	oo	o	o	oo	oo	o	o	ooo	ooo	ooo
External pavings	o	oo	ooo	ooo	oo	oo	oo	ooo	ooo	ooo
Cantilevered stairs	ooo	oo	o	o	ooo	o	ooo	ooo	ooo	ooo
Shelves	ooo	oo	ooo	ooo	ooo	oo	o	ooo	ooo	ooo
Rooves	ooo	o	oo	ooo	ooo	ooo	o	o	o	o
Bathroom furnishings	oo	o	o	ooo	o	o	o	o	o	o
Funerary art	o	o	oo	ooo	ooo	oo	o	o	o	o
Structural elements	ooo	ooo	ooo	ooo	ooo	o	oo	oo	oo	oo
ASTM Standard	C 97	C170		C97	C880/C99	D2845	C580/C1352	C241		
UNI Standard	9724/2	9724/3		9724/2	9724/5		9724/8	RD.2234	32.07.248.0	9724/6

0 = Not important; 00 = Important; 000 = Very important

or chemical elements in the material that can prejudice its use, either altering the esthetic characteristics recognized as valid by the market, or lowering physical-mechanical resistance far below the values typical of similar materials. Table 1.3 shows the most common defects found in ornamental stone materials.

FINE MARBLE IN ARCHITECTURE

Table 1.3: The Most Common Defects Found in Dimension Stone

TYPE OF DEFECT	SCIENTIFIC NAME	DESCRIPTION	MATERIALS AFFECTED	MAIN EFFECTS	POSSIBLE SOLUTIONS
Catena	Dike	Straight, light-colored veining from a few centimeters to a few dozen centimeters thick	Homogeneous and oriented granite	Aesthetic impairment	---
Biscie/Fili	Schlieren	Black, straight or undulating, thread-like veining, generally in bands a few dozen centimeters thick	Homogeneous and oriented granite	Aesthetic impairment	---
Patches/ Black spots	Mafic xenolith	Ovaloid, black body generally a few centimeters across	Homogeneous and veined granite	Aesthetic impairment	---
Oil stains		Greasy-looking, irregular-shaped rings a few dozen centimeters across	All types of black, green and white granite	Aesthetic impairment	Treatment of polished surface with solvents
Magrosità		Minute abrasions of the rock which cannot be polished; visible above all in glaring light	Marble and granite	Aesthetic impairment and higher probability of deterioration	Resin-coating of surface to be polished
Rust spots	Oxidation	Rust-colored stains and rings	Marble, granite and stone	Aesthetic impairment	---
Strappi	Microfractures	Minute fractures which appear as light-colored lines from a few centimeters to a few dozen cm long, often occurring in series	Marble and granite	Aesthetic impairment Possible reduction in the rock's mechanical resistance	Resin-coating of the polished surface
Durea	Silica concentrations	Irregular-shaped, grey or brownish-colored areas harder than the rocky mass which may have tiny cavities	Veined marble and granite	Aesthetic impairment	---
Tarolo	Cavities	Small, irregular-shaped holes, either sporadic or in series	Marble	Aesthetic impairment	Filling before polishing
Luccica	Glints from sparry calcite crystals	Much larger crystals than those in the surrounding rocky mass	Marble	Aesthetic impairment and structural damage in places	---
Frescume		Whitish- or yellowish-colored stripes or spots on the polished surface	Marble	Aesthetic impairment	---
Peli furbi	Microfractures	Closed fractures which open up during processing or after installation	Marble, granite and stone	Structural yielding	---
Peli ciechi	Microfractures	Fractures inside the block which are not visible on the surface	Marble, granite and stone	Reduction in block yield	---
Manine/ Stelline	Glints from crystalline surfaces	Light-colored, star-shaped reflections a few centimeters across	Absolute black granite	Aesthetic impairment	---

CHAPTER 2
MARBLE: THE PREMIUM MATERIAL

MARBLE PRIZED IN ANTIQUITY

The earliest evidence of stoneworking is found in Egypt, in the Saquara necropolis, specifically in the funerary complex of King Zoser, founder of the III Dynasty (2780 B.C.). This was a step pyramid made of limestone blocks designed by the vizier Imothep (considered to be the first known architect) and surrounded by a wall of well-shaped, well-sized limestone blocks. Inside the tomb a life-size statue of King Zoser in calcareous stone was found, attesting to the fact that, along with its use in building, stone was equally utilized for figurative works. Thus, stone became a prized material in Egypt with the use of alabaster as a decorative element, worked as slabs, objects, and sculpture. In Egypt this concept may have reached its highest interpretation when the red porphyry extracted from the famous Mons Porphyrites quarry became an authentic symbol of imperial power, taking on a value that lasted through Cleopatra's time (if it is true, as Lucan tells us, that the walls of her palace were not, as usual, clad with slabs but actually built of porphyry blocks).

It seems likely that the ancient Egyptians' concept of premium marble developed as a result of the high symbolic power attributed to the works created with it. According to some scholars, the dark colors, the surfaces that could be finely polished, and the hardness of stone that enabled works to last through time were perhaps associated with the idea of immortality and hence destined primarily for religious and monumental art. From this standpoint, materials of a silicate nature such as granite, syenite, and porphyry were probably the most prized since they could be used for works of not only great esthetic worth but of symbolic

Photo facing page: Detail of the façade of Santa Maria Novella, Florence. Leon Battista Alberti, circa 1440.

and commemorative value. The soft, nonpolishable materials were instead selected for the figurative arts in the strict sense, for works in any case enlivened with colorings and the addition of colored stones.

From the Egyptian courts the use of colored marble spread first to Greece and then became a common practice in Roman society. According to Pliny the Elder, the first important use of colored marble to cover walls was in the palace of Menander, satrap of Caria in the fourth century B.C. In ancient Greece the taste for colored marble also prevailed in sculpture. After the early, Bronze Age statues in white statuary marble found on the islands of Naxos and Paros—in which the white of the ground was an intrinsic part of the work's quality—it became common for color statues to have more expressiveness; some truly important works were covered with gold leaf.

In Rome the first testimonial to the use of colored marble dates from 149 B.C., when some of the paving of the temple of Capitoline Jupiter was laid. At that time, however, the use of marble was not yet common in the Roman world, and it spread only after Rome's acquisition of many active fine marble quarries in conjunction with its conquests of Greece and Carthage. Under Caesar and then Augustus, this taste for marble led to the opening of the Luni quarries (in what is now the Carrara marble basin) from which valuable Statuary was extracted in addition to other varieties.

Beginning in the first century A.D., marble use in the Roman Empire acquired a social significance, leading to greater differentiation between the use of common marbles and the use of colored or otherwise premium ones. White marbles were used mainly for cladding houses while colored marbles were more often destined for decorative flooring and facing inside public and private buildings. In the late Imperial period the distinction between common and premium varieties was even further accentuated: The first were used for funerary art (steles and epigraphs); the second for decorative purposes and to build exposed structural elements that had great esthetic value as well as architectural function. Prices obviously reflected these different market demands. In fact, according to Strabo, there was a great difference in price between white and colored marble and Diocletian's Edict shows that the costliest marble was red Egyptian porphyry while the cheapest was *proconnesius,* a white marble crossed with gray bands extracted on the island of Marmara. The demand for premium marble probably reached its height in the Flavian era when chromatic quality and rarity became fundamental factors in evaluating the worth of a stone material.

In addition to decorating floors and walls, colored marbles were often used in special, small-block works such as fountains, statues, pools, and objects of various kinds in which

FINE MARBLE IN ARCHITECTURE

the material's decorative function was well expressed. In time, stone's use as a covering took on an increasingly ornamental significance until slabs were cut just 2 mm thick, depriving them of any structural function. Incessant esthetic exploration also led to mirror-image installation, as we find in the Byzantine era (Santa Sophia in Constantinople and the Church of San Vitale in Ravenna) although it is quite likely this type of covering had been done even earlier.

But the use of marble, especially premium marble, in ancient Rome was not viewed by all as the spread of a positive esthetic model as in the other figurative arts. Pliny the Elder was a fervent believer in the futility of using marble to decorate homes. His aversion to it, as revealed in his *Historia Naturalis*, had first of all an *ante litteram* ecological basis, for he found it totally unjustified to destroy a mountain to get luxury objects from it. But the reasons for his contempt must obviously be found in the moral sense of his thinking, which was contrary to any form of luxury. He viewed highly prized and costly marbles as reprehensible symbols of decadence and, among other things, he believed they had a negative influence on the pictorial style of the time. In fact, at least at the beginning of marble's popularity, painting was used to imitate the colors and patterns most typical of marble coverings, providing decor less costly than stone and hence accessible to the less affluent classes. Later on, the appearance of mosaics led to utilizing marble to imitate painting, often in substitution of it, giving rise to what Pliny scornfully called *stone painting*.

With the end of the Roman Empire the use of marble in the Italic regions heavily declined and was oriented mainly towards an elite market consisting solely of ecclesiastical works and those commissioned by the nobility. Beginning in the fifth century A.D., imports of marble from the former imperial provinces and extraction from quarries in the peninsula—first of all the Luni—ceased entirely. The decline in supply led to a systematic dismantling of the marbles used to decorate the buildings of the Roman era. Ancient Rome became an authentic quarry from which to take *crustae*, columns, capitals, and anything else, to be reused as such but also to be cut and adapted to new use demands. The right to "excavate" Roman palaces was reserved to a chosen few, often to families of marble workers (*marmorai*) who were given Papal consent. These marble workers, generically called Cosmati (from the name of one of the stoneworking families of note in the first centuries of the second millennium) developed a new way of interpreting marble use, known as Cosmatesque art. The Cosmati's art was essentially expressed in church restoration and decoration, interpreting the Roman's *opus sectile* in a figurative context and in an architectural style typical of religious art. The materials chosen for Cosmatesque works, which were often church (thus, high-traffic) floors, frequently had to be vibrant in color and particularly resistant to wear. So the materials considered of greatest worth were above all colored granites and porphyry, while soft materials considered of greatest worth were above all colored granites and porphyry; soft marbles, although of great esthetic value, were used in very limited quantities.

During the Renaissance marble use changed, in line with the new conceptions that to different extents were influencing all forms of art. The setters of the new trends were no longer anonymous stone carvers but architects of the caliber of Raphael, Vignola, Ligorio who, leaving the rigid Cosmatesque scheme, created much more complex layouts of flooring and facing, doubtless closer to *opus sectile* than the Cosmatesque works. In the wake of these new expressive tendencies, in the mid-sixteenth century the art of marble *inlay* began, which consisted of creating geometric and figurative designs using pieces of differently colored marbles and juxtaposing them painstakingly in carving and inlay work. In its initial stages, even this new form of marble use was based exclusively on marbles from Roman ruins. Unlike the Cosmati, however, the new artists also utilized soft marbles, which were easy to work and available in the vast range of colors and shapes indispensable to the esthetic effect of the finished product. Polychrome breccias, alabaster, and cipollin attracted the market's interest again and demand was so great that lack of supply induced Cosimo dei Medici I—a true aficionado of inlaid marble (and the instigator of its development in Florence)—to search through his territories for new colored marbles. And as chance would have it, those territories were full of premium materials, and the period witnessed the start, or increased extraction of, marbles like Breccia Medicea, Giallo Siena, and Portoro—all materials of very high quality and unknown to the ancient Romans, who had ignored the peninsula's real extraction potential. Obviously, inlaid marbles were destined for works of great prestige, from church wall coverings to luxury furnishings. There was a particular taste for inlaid tables, true works of art created by artists like Vasari, Vignola, and Ammannati.

In the Renaissance, particularly with Michelangelo's work and following it, great interest was aroused in white statuary marbles as well, especially in the finest varieties of Carrara marble. These were selected on the basis of their structure and purity—viewed as the whiteness of the ground and an absence of veining—which made them particularly suited for sculpture.

From the Middle Ages to the late Renaissance, therefore, the worth of a marble was assessed on the basis of the mechanical properties and chromatic characteristics required by different styles or use destinations: high mechanical resistance for Cosmatesque art, lively colors, and easy workmanship for inlaid works. However, in neither case were these absolute values—often the material was only a tool to create the definitive work whose final value would be based on the artist's skill. Basically, the worth of a marble depended more on the final product made than on its intrinsic properties. These appraisal criteria, which led to evaluating marbles on the same level as those used for sculpture, underwent a gradual change that showed up fully in the eighteenth century. In any case, as early as the 1600s the appearance of new materials in quantities and sizes hard to obtain from Roman excavations had stimulated architects like Bernini to use marble for architectural and structural motives, reevaluating marble as a material with intrinsic qualities valuable in themselves.

In the eighteenth century a decisive return to mirror-image installation sanctioned the new use of marble per se, with its patterns and colors resulting from wholly natural and unique events. But between the eighteenth and the nineteenth centuries, the taste for marble and especially for rare and ancient marbles was taken even farther. In fact, marble aroused great interest as an antique, a relic of the past that was highly appreciated for its esthetic qualities and scientific value. Marbles worked in tiny tiles were gathered into collections called *studies*, with the pieces laid side by side so one could simultaneously admire the esthetic uniqueness of the individual variety and the overall appearance of the collection, which looked like a sort of natural polychrome mosaic. The esthetic effect and antiquarian interest was such that these marble studies were often used to adorn furniture tops, sills, consoles, and other pieces, passing from simple documentary collections to luxury objects of decor. In some cases, such as Ferdinando V's Villa Favorita in Naples, this use was extended to reconstructing the floors of Roman villas, reassembling the originals piece by piece.

In the nineteenth century the advent of Neoclassicism led to a new change in the taste for marble and hence in its value. A return to traditions carried with it a strong reassessment of white marbles, at the time coming from various parts of the world, to the detriment of colored ones, whose loss of appeal lasted until the mid-twentieth century.

In the mid-twentieth century, the premium marble concept was newly extended to colored marbles and later to siliceous materials, granites in the forefront. The discovery of materials with high mechanical resistance and patterns and colors similar to true marbles' was the right response to a growing need for materials suitable for exteriors supplied in qualitatively homogeneous large lots. A boom in large-scale projects soon overshadowed former interest in small production, perhaps qualitatively excellent but unsuited to big commissions.

In the last decade of the twentieth century, interest in premium marbles—following a trend that continues today—returned to valorizing colored and white statuary marbles and new materials like quartzite, even in some cases stones. Nowadays, the value of a material is based essentially on its appearance and its color. Mechanical characteristics may be partly overlooked, thereby relegating the material's real use potential to second place and suiting the use to the material, not vice-versa as was common in the past.

PREMIUM MARBLE TODAY

What do we mean today by premium marble? Is there a definition that describes its characteristics, and why is one marble considered finer than another? Is the concept univocal or does it derive from questions of taste, making it fairly ephemeral?

Actually, establishing criteria for defining materials that merit the term premium is quite a heavy responsibility. Even among insiders there are different opinions: The borderline between a premium marble and a "common" one may vary perceptibly between those working in marble, and they may, moreover, have convincing arguments to back up their own opinions. In all likelihood, the correct interpretation of a factor tied to the market can only derive—as often happens in the stone sector—from direct observation of the market itself.

We spoke before about the fundamental importance of a material's esthetic characteristics. In detail, the first aspect looked at in assessing a material's worth is usually its color. In general, pattern is an element subordinated to color but in some materials accentuates the chromatics, giving so much life and personality to the esthetic model that it becomes the main component. This is the case with many breccias but also with marbles in which the veining holds the chromatic element in dominating patterns that attract one's attention first.

Priorities aside, color and pattern often appear blended in an overall look that should be considered in its totality, including its possible (and in many cases frequent) variations. It is useful here to briefly digress into how perception of a premium material's qualities can also be influenced by objective factors.

Usually, a judgment about a material's esthetic value is made at the distance lying between the observer and the installed material. Only in this way does one have the right esthetic perception of the material, which can be compared to the market's taste. Now, unlike the marbles and granites used in large-scale coverings—where the installed material is always rather far from the observer—in the case of marbles used because they are prized, this distance is often reduced to just a few feet. This enables one to perceive the material's esthetic model in its tiniest details, from a basic to an overall qualitative assessment. In fact, we should emphasize that quite often the qualitative perception of a marble with premium characteristics varies even at short distances. For example, many monochromatic breccias have patterns that become invisible just a couple of meters away, and there are polychrome marbles which, if seen from afar, lose their color distinctions, changing the esthetic perception of the whole. Likewise, since the quality of premium materials is based mainly on color, the final esthetic effect will be heavily influenced by the amount of light in the place of installation. In the case of a dark marble, particularly a red or a green, slightly insufficient or even just badly distributed lighting could make it hard to fully appreciate the material's worth.

Another aspect often characterizing premium marbles is their availability in limited—if not downright scarce—amounts. In contrast to more common materials, the fact of limited availability not only has no influence on market potential but may even be a positive factor because it adds the charm of rarity to an already high-quality material. Naturally, scarcity also increases its economic value.

That the material may have an historical origin is a plus. The discovery that a material with prized esthetic features was also used in antiquity, perhaps for historically important works, can only increase its allure, adding historical value to its intrinsic beauty. The same may hold true for recently discovered materials that recall the forms and colors of famous varieties by now exhausted.

Finally, factors like block size, workability, and use possibilities, often of fundamental importance in common stone materials are, in the case of premium marbles, quite secondary. Certainly, specific limits have to be respected, below which the material is impossible to use, but the market has a high degree of technical and economic tolerance that tends to find work and use solutions allowing for use of the material in any case. This is especially evident in materials of high chromatic quality whose function, due to the meager quantities available, is often limited to decoration and to valorizing works done mostly with other materials.

Basically—in getting back to trying to define a premium marble—we could place in the category of premium marbles those whose esthetic features are so important they may modify the normal procedures of selecting, marketing, and using stone materials, and adapting them to their own technical characteristics, availability, and use. In other words, while a common marble must have physical-mechanical, production, and size prerequisites as well as esthetic ones to be a success on the market, the market itself adapts to the characteristics of a premium marble, within the limits of its workability as natural stone. And the more the market is willing to modify its own general standards, the more the marble will be prized.

In practice, esthetic characteristics of this caliber are found mainly in varieties belonging to the commercial group of marbles. Quite frequently these include breccias, either polychrome (*Brèche de Vendôme, Jasper, Breccia Capraia*) or practically monochrome (*Breccia Pernice, Rosso Carpazi, Fior di Pesco Apuano*), and the marbles whose colors have strong impact. Among the colors most admired and desired are blue (*Sodalite Blue Royal*), azure (*Azzurro d'Oriente, Azul Cielo*), black (*Nero Piemonte di Ormea, Black Pearl, Nero Belgio, Portoro*), green (*Verde Antigua, Ming Green*), yellow (*Giallo di Siena, Giallo Reale*), and red (*Rosso Collemandina, Rosso Verona*), always preferably in saturated tones, often of a certain intensity. White marbles, even the veined ones, have to have a clean, pale, bright ground (*Statuario, Bianco Acquabianca, Taxos White, Ambaji White*) or a warm ivory tone (*Statuario Venato, Calacata, Imperial White*) but always devoid of grayish tones. Pink is less frequent and has delicate tones, in which the color looks slightly shaded (*Rosa Aurora*).

In a class by themselves are the ophiolithic-type green marbles that may or may not have a brecciated pattern. Here the green is generally rather dark and frequently lightened by pale green veining (*Verde Patricia, Verde Rameggiato, Cavendish Antique, Verde Antico d'Oriente*) or white veins (*Verde Alpi*). Some of the veined marbles are distinctive for their decorative patterns, especially the green ones (*Cipollino Apuano, Cremo Tirreno*).

The quartzite group, while a small one, contains a good percentage of varieties with prized characteristics. The most frequent color is azure in various intensities (*Azul Macaubas, Quartzite Blu*) and in some cases with greenish and violet tones (*Azul Imperial*). There are also pink (*Pink Quartzite*) and reddish varieties.

The stone group includes varieties that differ greatly from one another but that usually do not have the characteristics to make them premium. Exceptions are some varieties with special chromatics (*Pietra Dorata*) and others that are considered premium less for their appearance than for their physical-mechanical properties, which allow them to be polished (*Pietra di Matraia*).

Among the granites we rarely find materials whose esthetics prevail over the other factors normally demanded of a stone material. This is probably due to two distinct causes: In

FINE MARBLE IN ARCHITECTURE

the first place, granite use is often large-scale, requiring large-scale production, homogeneous quality, and well-defined physical-mechanical properties. Second, save some worthy exceptions, the esthetic model for these materials never reaches the qualitative levels able to evoke that sense of the beautiful that so many marbles do. Obviously, there are some exceptions, the most outstanding of which are the blue granites (*Azul Bahia*).

CHAPTER 3
WORKING AND USING PREMIUM MARBLE

In the preceding chapter we tried to give a definition of premium marble on the basis of the market conditions that regulate commerce in these materials. Now let's look at the main types of workmanship and use.

Working premium marble is no different from working other ornamental stone materials but, unlike techniques for the latter, there are methods that enhance their esthetic characteristics. As already mentioned in regard to other parameters, the prized stone product is usually not subject to the use demands normally required for building purposes; rather, use fields derive from its quality.

Premium marble is most frequently worked in medium to large-size **slabs,** which enable the material's esthetic model to be admired in its entirety and valorize it with expressly studied use ideas. Above all, this is the case with marbles with a wavy pattern, which are more or less highlighted by a nonrepetitive polychromy whose physical continuity is often viewed by the designer as an element indispensable for the chosen use (fig.2). The most typical use is in mirror-sequence installations that either take a single direction (*open-book*) or two directions at right angles to one another (*mirror-image*) (fig. 3). Solutions of this type are always highly effective and often constitute the main (sometimes the only) element of the entire décor. They are utilized primarily in coverings of small and medium dimensions in which the mirroring effect is very visible and takes up a good part of the field of vision. They are also frequently used in flooring, although here the observation point is rather oblique and closer-to, which can limit the esthetic impact.

Premium marble is also often worked as **tiles** to exploit quarry production to the hilt

Photo facing page: Madonna with child and angels, Tomb of Filippo Strozzi, Santa Maria Novella, Florence. Benedetto da Maino, circa 1460.

FINE MARBLE IN ARCHITECTURE

fig. 2

fig. 3:
Cipollino Cremo Tirreno

fig. 4

by taking advantage of shapeless or highly fractured blocks. Moreover, this type of production can be especially opportune for materials whose esthetic model can be appreciated on limited surfaces with no detriment to its qualitative value. We refer in particular to materials with homogeneous color and pattern, to breccias without defined movement, and also to materials whose veining follows a slight movement appreciable even on a small scale.

In **cut-to-size** pieces for grid work, premium marbles (often combined with more common ones) are utilized primarily for their chromatic features, which make an essential contribution to the overall appearance of the work (figs. 4, 5, 6, and 7). In these cases premium

fig. 5

fig. 6

fig. 7

materials take on greater significance as decorative elements, usable with great profit even in small amount. Here preference is given to bright and deeply colored materials, perhaps for these very features unusable in large amounts.

Solid stone is also typical for premium marble because such pieces are often individual elements requiring great esthetic quality (figs. 8, 9). The most frequent items are vanity and table tops and shelving and counters. Especially where counters are concerned—pieces of a certain length—materials with a markedly oriented pattern are preferred and installed to run lengthwise on the manufacture. When the material is easy to work, curved pieces are also created to cover large-size columns.

Whatever the type of workmanship, the surface of a product made from premium marble is generally polished. Polishing is, in fact, the **finish** that best enhances the material's appearance and especially its color, saturating its tonalities. Not infrequently, however, especially if one wants to maintain the warm tones of certain pastels, fine-smoothing is preferable because it softens the marble effect and gives the surface a more natural look. An important variation on this surface is the antiqued, a sort of shaving treatment that gives the surface tiny imperfections such as scratches, dents, perhaps an accentuated *magrosità*, which overall reproduces the look and charm of a surface worn by time. Only in rare cases, and usually on stones, is flaming done to get a rough surface identical in color and irregularity to the material's nat-

fig. 8

FINE MARBLE IN ARCHITECTURE

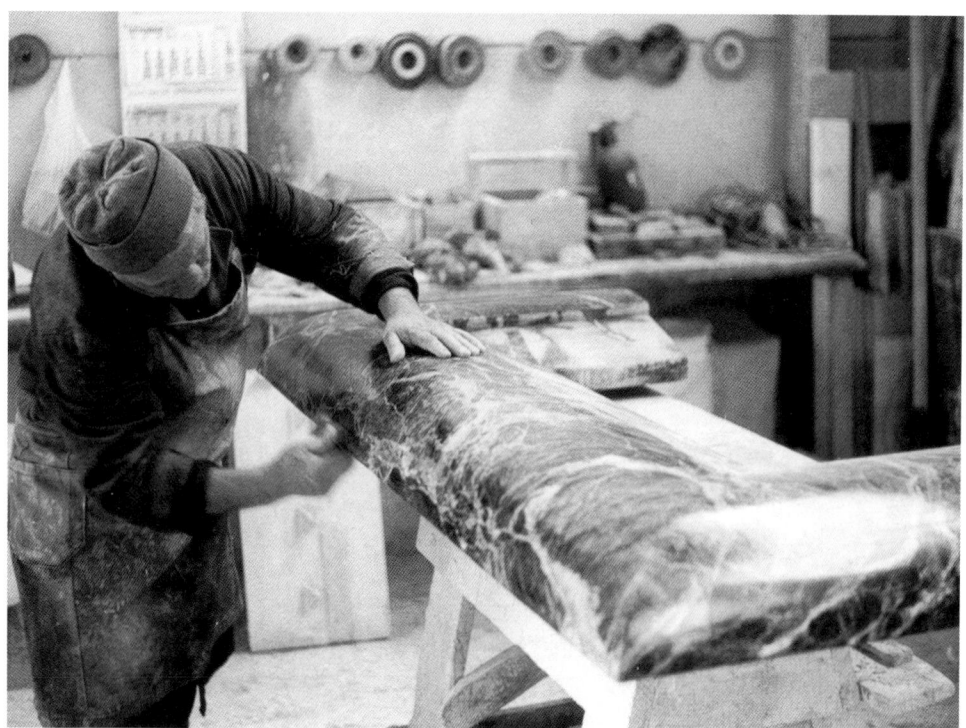

fig. 9

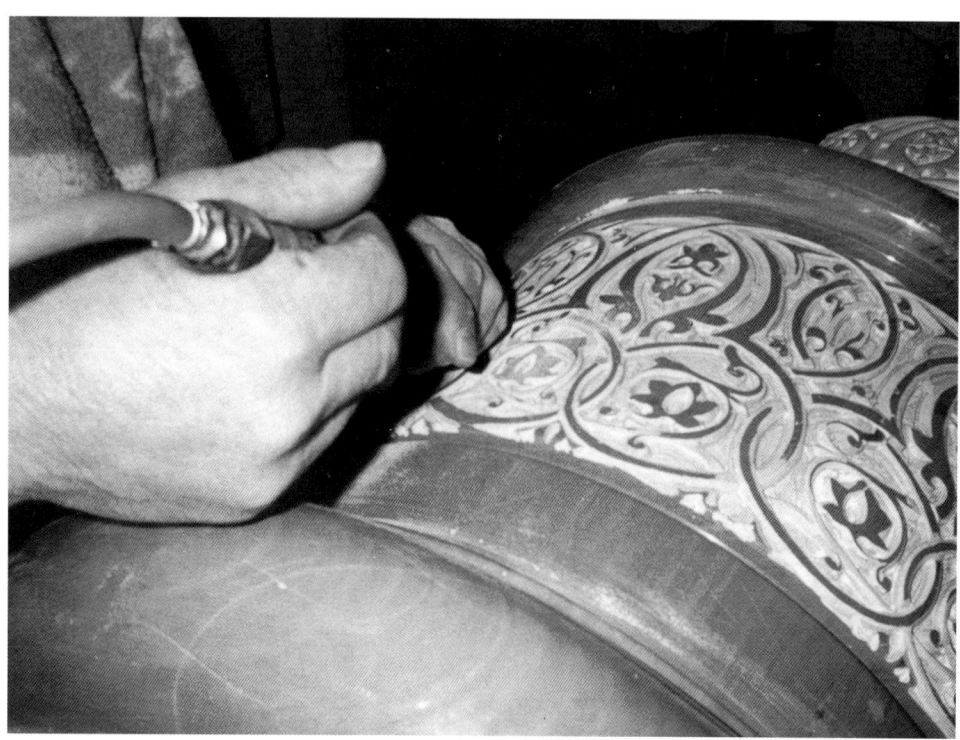

fig. 10

fig. 11

ural surface. Finally, given the esthetic purposes of premium marble, shock treatments in their various types are almost never given. In fact, percussion on a marble's surface has devastating effects on the material's color and pattern and generates a homogeneous surface without particular esthetic worth. However, shock treatments can be done on monochromatic materials in combination with another type of finish to create a pattern or get an especially effective color contrast (fig. 10).

On the **treatment** level, stone materials often need to be reinforced, either by coating with resin to fill in (fig. 11) and consolidate loose veins, *magrosità* or tiny cavities, or with plastic webbing on the rear of the slab or tile to give the finished product greater consistency and rigidity. Materials of particular worth but with marked structural weakness should be backed with a slab of common marble for support.

Green marbles of the ophiolithic type, given their special genesis and nature, often undergo perceptible stretching movements that cause breakage once they are cut into blocks or slabs, creating cracks that can reduce block yield or compromise the material's use. This phenomenon, found in other materials but not to the same extent, requires the blocks to be seasoned for some time before sawing so that the material can gradually release its inner tension without fracturing too much.

With regard to stones with a high absorption coefficient, it is often a good idea to coat

fig. 12: "Desiderio di Donna." Sculptor: Gualtieri

the visible side with an impregnating substance that preserves the material's esthetic and physical-mechanical characteristics.

In addition to their use in building—undoubtedly their biggest use field—premium marbles are also used in sculpture. Following the Neoclassical trend the materials most commonly used for the human figure are white statuary marbles, either in absolutely pure (fig. 12) or warmer tones that tend to ivory. For abstract works the choice of materials instead falls

FINE MARBLE IN ARCHITECTURE

to a vast range of colored marbles, especially those with uniform colors or only slight variations in tone. These are mainly black, red, and yellow marbles, often devoid of veining or at most with a light design. Exceptions to this are some important varieties of blue marbles and quartzites in which, on the contrary, shadings of tone and chromatic depth (often combined with decisive veining) are the esthetic models of the artist's choice.

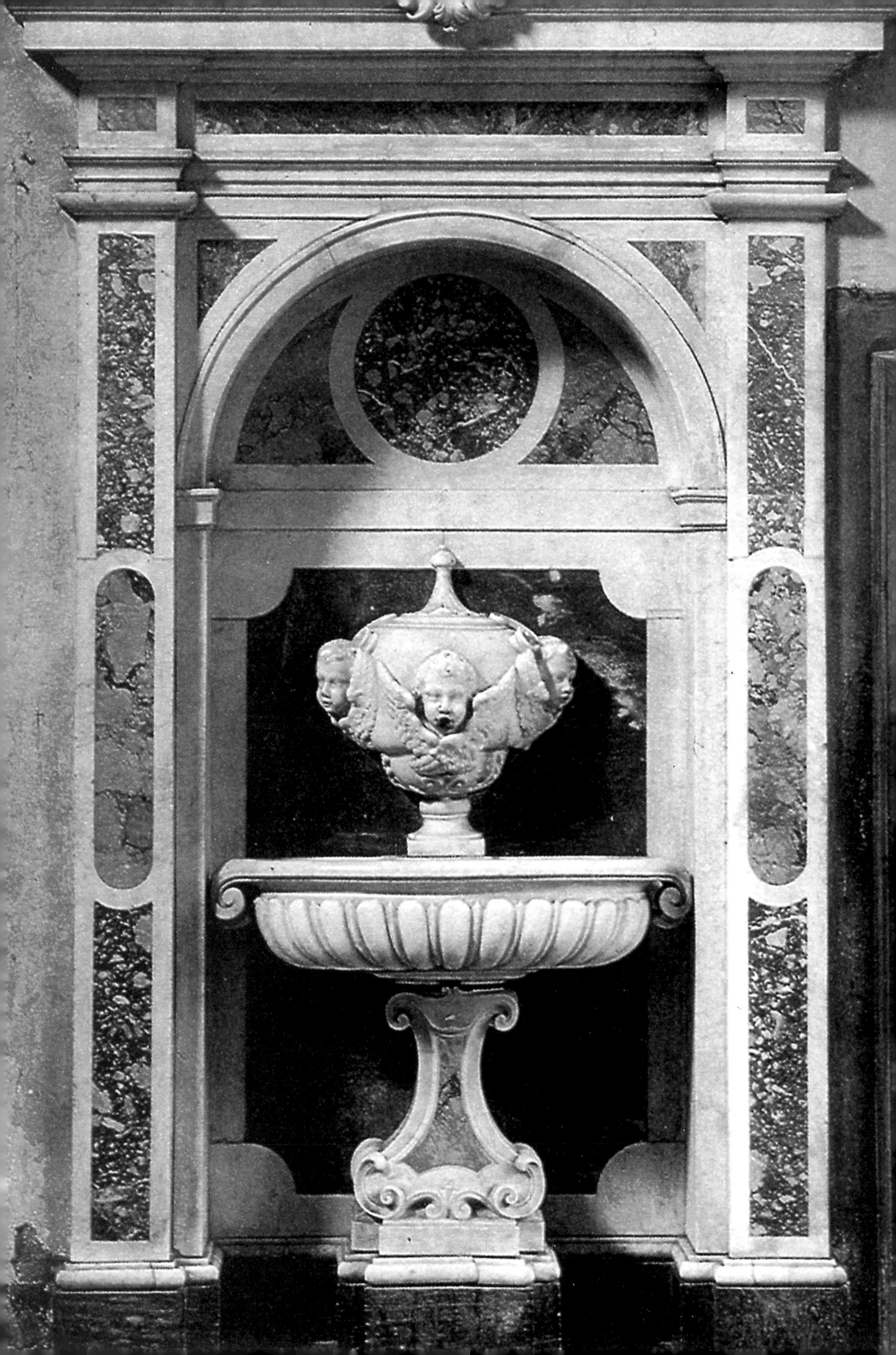

CHAPTER 4
PHYSICAL-MECHANICAL DATA

The following pages contain the results of physical-mechanical tests run on some of the stone materials shown in Part II. The values were furnished by producers or suppliers or obtained from books. For the other materials dealt with in the book but not included in the following list, it appears no physical-mechanical tests have been done.

DESCRIPTION OF TERMS

Weight per unit of volume = the ratio between the weight and the apparent volume of the material (i.e., including the holes inside the material too or its porosity) which corresponds to its actual weight. This is fundamental in the assessment of the loads that a material exerts on the fixing and support structures. It also provides information on the compactness of the material.

Imbibition coefficient = a material's capacity to absorb liquids—this gives an indication to its porosity. It is important to know this if the material is to be used where it is frequently in contact with liquids (bathrooms, kitchens, exterior uses, etc.)

Compression breaking load = a material's ability to withstand loads. This is important for materials that are made into products with a structural function (columns, arches, etc.) but it is also important for cladding and flooring, which, for many reasons, are subjected to compressive stress.

Compression breaking load after freezing = a material's ability to withstand loads in environments with significant temperature changes above and below zero degrees Celsius. It is

Facing page: Lavabo, Santa Maria Novella, Florence. Giovacchino Fortini.

fundamental to have this information if a material as to be used externally (paving, cladding) in climatic conditions with significant temperature changes over 24 hours.

Flexural strength = a material's ability to withstand bending pressure. This is extremely important because mechanical stress causes bending either directly or indirectly in many applications.

Frictional wear test = the material's ability to withstand wear. This is important in assessing a material's suitability for internal and external flooring, shelves, etc.

MPa = MegaPascal.

N/mm^2 = Newton over squared millimeters.

Ambaji White

Weight per unit of volume	2720	kg/m³
Imbibition coefficient	0.064	%
Compression breaking load	78	MPa
Compression breaking load after freezing	73	MPa
Flexural strength	8.8	MPa
Frictional wear test	6.9	mm

Arabescato Piana

Weight per unit of volume	2675	kg/m³
Imbibition coefficient	0.14	%
Compression breaking load	1265	kg/cm²
Compression breaking load after freezing	1200	kg/cm²
Flexural strength	112	kg/cm²
Frictional wear test	5.02	mm

Azul Imperial

Weight per unit of volume	2680	kg/m³
Imbibition coefficient	0.4	%
Compression breaking load	117	N/mm²

Azul Macaubas

Weight per unit of volume	2683	kg/m³
Imbibition coefficient	0.11	%
Compression breaking load ear test	210.2	MPa
Compression breaking load after freezing	212.9	MPa
Flexural strength	20.39	MPa
Frictional wear test	0.54	mm

Black Pearl

Weight per unit of volume	2740	kg/m³
Imbibition coefficient	0.036	%
Compression breaking load	137.3	MPa
Compression breaking load after freezing	135	MPa
Flexural strength	28	MPa
Frictional wear test	4.17	mm

Breccia Pernice

Weight per unit of volume	2640	kg/m³
Imbibition coefficient	0.4	%
Compression breaking load	2145	kg/cm²
Frictional wear test	0.28	mm

Breche de Vendôme

Weight per unit of volume	2717	kg/m³
Compression breaking load	109	N/mm²
Compression breaking load after freezing	98	N/mm²
Flexural strength	12	N/mm²
Frictional wear test	4.49	mm

Calacata Borghini

Weight per unit of volume	2695	kg/m³
Imbibition coefficient	0.13	%
Compression breaking load	1235	kg/cm²
Compression breaking load after freezing	1135	kg/cm²
Flexural strength	194	kg/cm²

Cipollino Apuano

Weight per unit of volume	2704.8	Kg/m³
Compression breaking load	885	MPa
Compression breaking load after freezing	797	MPa
Flexural strength	83.7	MPa
Imbibition coefficient	0.21	%

Cipollino Cremo Tirreno

Weight per unit of volume	2719	Kg/m³
Compression breaking load	1247	MPa
Compression breaking load after freezing	1074	MPa
Flexural strength	146.2	MPa
Imbibition coefficient	0.13	%

Diaspro Tenero di Sicilia

Weight per unit of volume	2599	kg/m³
Compression breaking load	137	N/mm²
Compression breaking load after freezing	87	N/mm²
Flexural strength	14	N/mm²
Frictional wear test	4.6	mm

Giallo Reale

Weight per unit of volume	2695	kg/m³
Imbibition coefficient	0.14	%
Compression breaking load	1773	kg/cm²
Compression breaking load after freezing	1831	kg/cm²
Flexural strength	125	kg/cm²

Giallo Siena

Weight per unit of volume	2710	kg/m³
Imbibition coefficient	0.35	%
Compression breaking load	1783	kg/cm²
Compression breaking load after freezing	1656	kg/cm²
Flexural strength	259	kg/cm²

Imperial Danby

Imbibition coefficient	0.16	%
Weight per unit of volume	2723	kg/m³
Compression breaking load	68.21	MPa
Flexural strength	10.97	MPa

Nero Piemonte di Ormea

Weight per unit of volume	2689	kg/m³
Imbibition coefficient	0.3	%
Compression breaking load	1089	kg/cm²
Compression breaking load after freezing	979	kg/cm²
Flexural strength	209	kg/cm²

Noir Saint Laurent

Weight per unit of volume	2700	kg/m³
Compression breaking load	127	MPa
Frictional wear test	24	mm

Pietra di Matraia

Weight per unit of volume	2664	kg/m³
Imbibition coefficient	0.25	%
Compression breaking load	1869	kg/cm²
Compression breaking load after freezing	1759	kg/cm²
Flexural strength	382	kg/cm²

Pietra Dorata

Imbibition coefficient	3.14	%
Compression breaking load	51	MPa
Compression breaking load after freezing	45	MPa
Flexural strength	8.9	MPa
Frictional wear test	7.38	mm

Portoro

Weight per unit of volume	2712	kg/m³
Imbibition coefficient	0.08	%
Compression breaking load	1862	kg/cm²
Compression breaking load after freezing	1597	kg/cm²
Flexural strength	104	kg/cm²

Quarzite Blu

Imbibition coefficient	0.14	%
Compression breaking load	235	MPa
Flexural strength	10.7	MPa

Rosso Carpazi

Weight per unit of volume	2760	kg/m^3
Imbibition coefficient	0.18	%
Compression breaking load	1424	kg/cm^2

Rosso Verona

Weight per unit of volume	2691	kg/m^3
Imbibition coefficient	0.18	%
Compression breaking load	1626	kg/cm^2
Compression breaking load after freezing	1531	kg/cm^2
Flexural strength	100	kg/cm^2

Rouge France Incarnat

Weight per unit of volume	2700	kg/m^3
Compression breaking load	111	MPa
Frictional wear test	24.1	mm

Statuario

Weight per unit of volume	2.688	kg/m^3
Imbibition coefficient	0.1	%
Compression breaking load	1.235	kg/cm^2
Compression breaking load after freezing	1.194	kg/cm^2
Flexural strength	187	kg/cm^2
Frictional wear test	5.3	mm

Travertino Rosso

Weight per unit of volume	2596	kg/m³
Imbibition coefficient	1.15	%
Compression breaking load	529	kg/cm²
Compression breaking load after freezing	514	kg/cm²
Flexural strength	111.2	kg/cm²
Frictional wear test	4.2	mm

Verde Patricia

Weight per unit of volume	173.01	Lbs/ft³
Imbibition coefficient	0.29	%
Compression breaking load	21.915	psi
Flexural strength	2701	psi
Frictional wear test	37.79	hardness

Verde Rameggiato

Imbibition coefficient	0.22	%
Weight per unit of volume	2622	kg/m³
Compression breaking load	150	MPa

Details of facede of Santa Maria Novella, Florence

Base of the altar, Cappella dei Bardi, 18th century

PART II
THE MARBLES

ESTHETIC CHARACTERISTICS

Ambaji White's appearance recalls the classic model of premium marble par excellence: white, often without veining, and thus without pattern, and chromatically homogeneous. Short, discontinuous dark green veins stand out here and there from the surrounding white ground like little decorative elements. Grain is medium-fine.

PRODUCTION

Ambaji White is extracted in the state of Gujarat in India. The quarries are located near the town of Ambaji, from which the marble gets its name, and have considerable output: over 2,000 m^3 a month in squared blocks sized 200 x 100 x 100 cm.

WORKMANSHIP AND USES

Ambaji White can be worked in all ways suitable for ornamental stone. It is available in tiles and small slabs, either polished or smoothed. Uses are primarily flooring and facing, although it is frequently worked as solid stone for columns and tops, as sculpture, and for special works made to measure. The material can be utilized outdoors as well and is not harmed by sunlight.

HISTORY

Ambaji White has a long history. For over a thousand years it has been used in India for temples and monuments and was often chosen by reigning princes for their private dwellings. Among the most famous works is the Delwara temple built 1,200 years ago.

WORKS

The most important recent works created with Ambaji White include the Ambaji Temple and the Swaminarayan Temple, both in Neasdon, near London.

THE COMPANY

Ambaji White has been produced and marketed by Trivedi Crafts Pvt Ltd. since 1937. The company was created to exploit the Ambaji quarries in order to rebuild the area's ancient temples and is now a leading part of the Trivedi Marble Group. The group's work runs from extracting to supplying finished products, including special works. It also owns a Black Pearl (Abu Black) quarry and another of Rajasthan Green. It has a tile plant in Ambaji, a slab plant in Ahmedabad, and a plant in Ahmedabad for special works (carved products). Thanks to its up-to-date architecture and design division, the company can realize projects for any type of commission.

Trivedi Crafts Pvt Ltd.
68 Premanjali Society, Bodakdev, Ahmedabad 380054 (India)
ph. +91 79 674 5522/3110/4510; fax +91 79 676 6220;
Internet: http://www.trivedi-marble.com; e-mail: trivedi_kiran@vsnl.com

Ambaji White

ESTHETIC CHARACTERISTICS

Bianco Acquabianca has a bright white fine-grain ground; it has a uniform purity, with neither veining nor shadings in a felicitous synthesis of simplicity and rarity typical of a stylish material above the dictates of fashion. This is the esthetic picture of the finest variety of Bianco Acquabianca, fully corresponding in quality and features to the better known Bianco P.

PRODUCTION

Bianco Acquabianca is extracted from several quarries in the homonymous basin on the northern slope of the Apuan Alps in the community of Municciano (Lucca). Total annual production is about 6,000 to 8,000 tons and about 30 percent is in squared blocks sized an average 300 x 150 x 160 cm. The rest is in semi-squared and shapeless blocks.

WORKMANSHIP AND USES

The main use of Bianco Acquabianca is as statuary in general and in funerary works. It is also frequently used in small blocks for columns and other special pieces, while utilization in slabs and tiles is much less common. The material can be used in exteriors.

HISTORY

Extraction in the Acquabianca basin dates from early last century.

WORKS

Bianco Acquabianca has been used in many prestigious projects worldwide. Among the most significant are the J. F. Kennedy Mausoleum in New York City and American military cemeteries in Greve in Chianti (near Florence) and in Honolulu. A more recent work is the Sultan Qaboos Mosque in Oman.

THE COMPANY

Bianco Acquabianca is extracted by Coop. Bianco Campaccio s.r.l. Constituted in 1989, the cooperative took over four quarries from Imeg, which also produce Bianco P., Bianco Campaccio, Bianco Scaglia, Bardiglio Imperiale, Bardiglio Ordinario, and Arabescato Carcaraia. The company directly handles sales of its own materials.

Coop. Bianco Campaccio s.r.l.
Piazza Pancetti 5, 55030 Gorfigliano-Minucciano (LU) Italy
ph. +39 0583 610016; fax +39 0583 610016
Stock Yard: Via Aurelia 100, 54033 Avenza Carrara (MS) Italy
ph. +39 0335 262424

Bianco Acquabianca

ESTHETIC CHARACTERISTICS

Imperial Danby has a warm white ground with slight movement by a wavy, discontinuous hatch of short gray and golden shadings like small, sparse brush-strokes, just visible enough to give its pattern a delicate orientation. It is medium-grained.

PRODUCTION

It is the finest variety extracted from the Danby quarry near Rutland, Vermont. This quarry, probably the world's largest underground marble quarry, has a total monthly output of about 1,200 tons, half of which consists of Imperial Danby. Blocks are generally medium-sized.

WORKMANSHIP AND USES

Imperial Danby is worked in tiles and medium-sized slabs for flooring and facing, paving and cladding. It is very frequently worked in small blocks for columns, fireplaces, tops, and special jobs in general. It can be given polishing, smoothing, and shock treatments.

HISTORY

The Danby quarry was opened in the early twentieth century.

WORKS

Among the major works created with Imperial Danby are prestigious projects like the Supreme Court Building and the Thomas Jefferson Memorial in Washington, D.C.

THE COMPANY

Imperial Danby is extracted by the Vermont Quarry Corp., an American company founded in 1992 by R.E.D. Graniti s.r.l. along with another Italian concern. The R.E.D. Graniti Group is a world leader in excavating and marketing granite and marble in blocks, and it offers famous materials like Giallo Veneziano (exclusive), Verde Eucalipto, Nero Africa, Blue Pearl, Baltic Brown, Balmoral Red, and other red Finnish granites.

R.E.D. Graniti s.r.l.
Via Dorsale 12, 54100 Massa (MS) Italy
ph. +39 0585 88471; fax +39 0585 884848
Internet: www.redgraniti.com; e-mail: info@redgraniti.com

Imperial Danby

ESTHETIC CHARACTERISTICS

Statuario Michelangelo's special feature is its warm ivory ground, which clearly distinguishes it from other Apuan materials. In places, the ground mass is crossed by shaded veins of cream tending to golden. Its fine grain gives the material particular consistency. When the veining thickens and becomes more frequent, forming a brecciated pattern, it takes the name of Calacata Michelangelo.

PRODUCTION

Statuario Michelangelo is extracted in the Polvaccio quarry in the Carrara marble basin. The quarry is located in the Torano zone, long known as one of the richest in fine marbles of the entire Apuan chain. And as is usual for many premium materials, at Polvaccio, too, this statuary marble does not constitute the entire quarry but outcrops discontinuously, mixed with Bianco Carrara. So production runs in alternate phases of scarcity and abundance, depending on the quarry fronts being worked.

WORKMANSHIP AND USES

As its name indicates, the use of Statuario Michelangelo par excellence is in statues. In sculpture it finds its highest expression especially when devoid of veins, because this means both no imperfections and because its structure allows for its carving. It is also frequently used in slabs and, to a lesser extent, in tiles for flooring and facing. Its use outdoors is uncommon, partly due to its high cost, making it a material to be utilized in limited quantities. The surface is polished or fine-smoothed.

HISTORY

Many believe that the Polvaccio quarry from which Statuario Michelangelo comes is the same that supplied Michelangelo with many of the marbles he used for his sculptures, first among them his Pietà. The quarry—active in Roman times—still shows many traces of ancient excavation, and it seems that it was active during the Renaissance.

WORKS

Statuario Michelangelo has been used for a great many works, particularly in luxury dwellings and in art works in the widest sense. An especially representative example is the altar in the new presbytery of the Padua Cathedral, designed by Giuliano Vangi.

THE COMPANY

Cave Michelangelo s.r.l. was founded in 1987 on the initiative of several companies working in the stone sector for many years and is currently one of the most important producers of statuary marble. The company works mainly in extracting and selling the marble in blocks and slabs and can offer any type of finished product. Among its major commissions for raw or finished products were the Buddha Sothorn Grand Royal Temple in Thailand and Place Charles de Gaulle in Vichy, France (both in Bianco Polvaccio), and the statues of Santa Brigida and Santa Caterina da Siena in the walls enclosing Vatican City.

Cave Michelangelo s.r.l.
Via Piave 32, 54033 Carrara (MS)-Italy
ph. +39 0585 842496 / 841408; fax +39 0585 845090 / 842957
e-mail: cave_michelangelo@tin.it

Statuario Michelangelo

ESTHETIC CHARACTERISTICS

The uniqueness of Arabescato Piana lies in its pattern, a more or less close-knit network of clear-cut, marked veining standing out from a clean white background that resembles statuary white in places. The veins often run straight, sometimes zigzag, and weave together in no specific motif, leaving room for random shapes. To be noted is the absence of taroli, the minute cavities that often depreciate Carrara marbles.

PRODUCTION

Arabescato Piana is extracted in just one quarry in the Colonnata marble basin (Carrara, Italy), located in "La Piana." The quarry, worked as both pit and tunnel, produces a total of 1,000 tons of marble, 300 of which are Arabescato Piana in blocks sized 300 x 150 x 150 cm. The rest of output consists of Bianco Piana, a special variety of Bianco Venato Carrara.

WORKMANSHIP AND USES

Thanks to its particular design, Arabescato Piana can be worked both in slabs and in 30 x 30 cm tiles without jeopardizing its esthetic features. When worked in slabs, mirror-image applications are quite effective. It is most frequently used internally as flooring and facing but has also been utilized as cladding. Many designers choose it for special small-block works such as sills, curbing, and cantilevered stairs.

WORKS

Arabescato Piana has been used worldwide for over half a century. Among the most important recent works using this material are the interiors of China Telecom and the cladding on the People's Insurance Building, both in Shontou, China. It has also been used extensively in private homes.

THE COMPANY

Arabescato Piana is extracted exclusively by Mirko Menconi Marmi s.r.l., in its own quarry which also produces Bianco Piana. Along with extraction work, since 1966 the company (founded in 1925) has also been processing its output. In the early 1970s it was the first company in the world to make 300 x 150 x 7 cm facing tiles, thus opening an entirely new market. Today, Mirko Menconi Marmi has a technologically vanguard, full-cycle processing plant for white and colored marble that can turn out slabs and tiles in standard formats as well as made-to-measure.

Mirko Menconi Marmi s.r.l.
Via Del Bravo 18, 54031 Carrara Avenza (MS) Italy
ph. +39-0585-857848/9; fax +39-0585-857847
Internet: http://www.mirkomenconimarmi.it; e-mail: mmm@mirkomencomarmi.it

Arabescato Piana

ESTHETIC CHARACTERISTICS

On a warm white ground, often tending to the color of statuary, Breccia Capraia has a pattern of veining that is alternately purplish red, iron gray, and greenish. Where the veining is denser and thicker, the marble takes on the typical brecciated look in which the elements elongate in a definite direction, giving the pattern movement and continuity. On the other hand, if the vein is thinner or sparser, this attenuates the shapes of the individual elements and becomes a characteristic stretch of color, uncertain and sinuous in its course and highlighted on the white ground.

PRODUCTION

Breccia Capraia is extracted from a single quarry located on the Tyrrhenean side of the Apuan Alps in Tuscany. The quarry, worked underground, produces other premium marbles such as Fior di Pesco Classico and materials like Statuario, Arabescato, Calacata, and Cremo, which are all comparable in quality to the better known Apuan varieties with the same names. About 6,000 to 7,000 tons of Breccia Capraia are extracted yearly in large squared blocks, and apropos of this it should be mentioned that the particular integrity of the deposit often makes it possible to get commercially outsized blocks destined for special works.

WORKMANSHIP AND USES

Worked primarily in slabs, Breccia Capraia is mainly used for facing and flooring in luxury interiors. Its brecciated, veined pattern lends itself well to mirror-image applications, and it is also frequently used for tops and special dimension work.

HISTORY

The epoch in which the quarry was opened is not known, but probably dates from at least the end of the nineteenth century. Attesting to this is the old sled route used to manually slide the blocks downhill, traces of which are still partly visible along today's access road to the quarry.

WORKS

Among the most significant works done with Breccia Capraia are the flooring and facing inside the Home Savings of America Tower in Los Angeles; the Wasserturm in Freiburg, Germany; and the Schwarzemberg Hotel in Vienna.

THE COMPANY

Breccia Capraia is extracted and sold by E.B.C. of Massa. The company, founded in the early 1970s, is part of the GMC Group which includes three other concerns with premises in, respectively, Ortonovo, Massa, and Porto Torres. The group also owns a Bianco Carrara quarry and two of Sardinian granite (Rosa Beta and Rosa Antico). In its own plants the group goes through the entire production cycle, from extraction to the finished product, working both marble and granite. Among its major commissions were the New China Building in Guangzhou, China; the Taiwan Cement Building in Taipei; the new Communist Party headquarters in Shanghai; and the Pulawska Financial Center in Warsaw.

GMC di Grassi Luciano & C. S.p.A.
Via Fossone Basso 8, 19034 Isola di Ortonovo (SP) Italy
ph. +39 0187 661680; fax: +39 0187 66350
e-mail: info@gmcspa.com

Breccia Capraia

ESTHETIC CHARACTERISTICS

Calacata Borghini is distinctive for a warm-toned white ground tending slightly to ivory, crossed by tobacco-colored veining shading into gray. The veining may be threadlike or banded and from a single element may thicken to form a brecciated pattern. In this case the material takes on movement that is lively and delicate at the same time. When the veining is sparser and the ivory ground predominates it is known as Statuario Borghini.

PRODUCTION

Calacata Borghini is extracted in Carrara in a quarry located in the classic Calacata zone (on the crest of Crestola in the Pescina-Boccanaglia basin). The quarry's selected production, including Statuario Borghini, amounts to about 600 to 700 tons a month both in squared blocks averaging 300 x 160 x 160 cm and semi-squared and shapeless blocks.

WORKMANSHIP AND USES

Calacata Borghini is worked in slabs and tiles for flooring and facing, paving and cladding. When the veining is quite evident interesting esthetic effects can be obtained in mirror-image installations. Surface treatments are polishing or fine-smoothing; the second gives the material an attractive antiqued look. Also frequent are small-block products for special works, for inlays and for quality fixtures like columns, fireplaces, etc. The material can be reinforced with epoxy resin without detriment to its quality.

HISTORY

Calacata has been extracted in the Crestola zone for a very long time, as attested to by the many manual cuts and old manufactures found in the extractive sites still operating. It is quite possible that the area was already known and exploited as early as ancient Roman times.

WORKS

Calacata Borghini has been chosen for numerous hotels in the Ritz Carlton chain. Other important uses have been the Twin Towers in New York, the Hyatt Hotel in Hong Kong, and the Lowe Hotel in Monte Carlo.

THE COMPANIES

Calacata Borghini is extracted by Calacata Crestola s.r.l. and marketed as finished products by Imarmi s.r.l. Both companies belong to the Marbo Group of Carrara.

Marbo di Borghini Paolo & Co. s.r.l.
Via Roma 13, 54033 Carrara (MS) Italy
ph. +39 0585 72993; fax +39 0585 73878

Imarmi s.r.l.
Via Carriona 230, 54033 Carrara (MS) Italy
ph. +39 0585 842671 / 846590; fax +39 0585 847282

Calacata Borghini

ESTHETIC CHARACTERISTICS

Paonazzo has a statuary-white ground of a warm ivory tone which, along the elegant veins crossing it randomly, deepens in color to form a yellowish halo. The veining, with clear-cut edges and a well-defined run, may be purplish red but is more often dark green and sometimes tends to grayish. In places the red vein widens into spots, giving Paonazzo the typical chromatic feature from which it takes its name. Not infrequently tiny pyrite crystals emerge here and there on the ground where they form delicate yellowish dots, a rusty look that increases the material's esthetic value.

PRODUCTION

Although Paonazzo has fair extraction potential, its output suffers from a lack of business promotion. It forms a continuous level in some Statuario marble quarries in the Carrara marble basins and is currently extracted solely as a byproduct of that variety. In any case, at least one quarry intends to start its active extraction quite soon. Blocks are squared and medium-sized to large.

WORKMANSHIP AND USES

Paonazzo can be worked in slabs or tiles without modifying its esthetics. It is a material to be used in interiors, generally as facing and flooring in combination with other materials. It is very interesting when worked in small blocks for vanity tops and bathroom fixtures. The surface finish is always polished. The material may need to be resin-coated to reinforce it.

HISTORY AND WORKS

Paonazzo was extracted more in the past than today. There is no precise date for its appearance on the market but in all likelihood can be traced to at least the late nineteenth century. Among recent works, we mention the interiors of the White House in Washington D.C. and the restoration of some parts of Harrod's in London.

THE COMPANIES

Paonazzo is extracted by the Aldo Vanelli Marmi di Giorgio Vanelli company, one of the most important producers of Statuario. The quarry, in operation for a great many years, has belonged to the Vanelli family since 1936 and also produces Venatino Betogli and Statuarietto. Output is handled exclusively by another family-owned company, Statuaria Marmi s.r.l., which markets it internationally as raw and finished product.

Aldo Vanelli Marmi di Giorgio Vanelli
Statuaria Marmi s.r.l.
Via Ilice 15, 54033 Nazzano Carrara (MS) Italy
ph. +39 0585 846300; fax +39 0585 846300
Internet: www.statuariamarmi.it; e-mail: info@statuariamarmi.it

Paonazzo

ESTHETIC CHARACTERISTICS

Calacata Luccicoso is characterized by shadings of a clear hazel color running to tobacco and pale gray, which is distributed over a warm white ground. With their shape and in their straight and diagonal runs the shadings are not isolated elements but often go from one to another uninterruptedly, covering a large part of the white ground or forming ovoid and elongated bodies on it. Sometimes minute pyrite crystals may be found. The grain is fine.

PRODUCTION

Calacata Luccicoso is extracted from an underground quarry in the Massa marble basin in the Apuan Alps. Output is about 300 tons a month in semi-squared and squared blocks running from 200 x 100 x 100 cm to 300 x 200 x 200 cm.

WORKMANSHIP AND USES

Calacata Luccicoso is basically worked in polished tiles and slabs but is also often smoothed. It is important that sawing follow the grain to highlight the diagonal shadings. The most frequent uses are facing and flooring and only interior use is recommended.

WORKS

One of the most prestigious projects created with Calacata Luccicoso is One Market Plaza in San Francisco, designed by Cesar Pelli & Associates.

THE COMPANIES

SO.LU.BER s.r.l. handles most Calacata Luccicoso production and is the main international supplier of the raw product. Founded in 1983, the company produces slabs of white and colored marbles and stones in general and can also furnish finished products. Among its major commissions were raw and finished Calacata Luccicoso for the Al Hani Tower in the United Arab Emirates. Carli Cav. Oreste & C. s.a.s. realized the work we show in this volume, one of the most interesting uses to which Calacata Luccicoso has been put. The company handles the full production cycle for marble and granite, from sawing to made-to-measure finished products. In addition to the aforementioned One Market Plaza, major works include the Ana Hotel in San Francisco; the Oracle Campus in Redwood City, California; and the San Jose Museum of Arts.

SO.LU.BER. s.r.l.
Viale Puccini 30, 54100 Massa (MS) Italy
ph. +39 0585 489487; fax +39 0585 45760

CARLI CAV. ORESTE & C. s.a.s.
Viale D. Zaccagna 47, 54031 Carrara Avenza (MS) Italy
ph. +39 0585 633343; fax +39 0585 631185

Calacata Luccicoso

ESTHETIC CHARACTERISTICS

Calacata Sponda is a material which, while a member of the Calacata family in the widest sense, is distinctive for its cream-colored ground, which shades to ivory. The veining, gray with beige shadings, can vary in thickness and density but never prevails over the ground mass, thus giving the material an overall look of basic homogeneity. Grain is fine. In places the veins may be thicker and the ground may show vague shadings of yellow.

PRODUCTION

The company whose works with Calacata Sponda we mention here does not actually quarry the material. Calacata Sponda is extracted in the Torano basin in a quarry in the Sponda locality famous for its inclusion in the classic area producing calacata-type materials. The quarry also turns out a small amount of Bianco Statuario. Calacata production is about 2,000 tons a month, 80 percent in squared blocks and 20 percent in semi-squared blocks. In both cases the average size is 270 x 150 x 100 cm.

WORKMANSHIP AND USES

Calacata Sponda is essentially worked in slabs and tiles for flooring and facing. Outdoor use can generate oxidation processes that damage the material's appearance. It is also used in small blocks. The finishing treatment is polishing or fine smoothing, the latter advisable for high-traffic floors or when an antiqued look is desired. Calacata Sponda requires no reinforcement or consolidation treatments.

HISTORY

Active since the Renaissance, the Sponda quarry mainly produced Bianco Statuario. Recently, for technical reasons, production has turned mainly to Calacata, a material that seems not to have been used in past epochs.

WORKS

Two of the most significant are the Brent Cross Commercial Centre in London and the Liffey Valley Mall in Dublin. In both cases Calacata Sponda was used in smoothed 30 x 30 cm floor tiles and in small blocks for sills, architraves, etc. International Italmarmi di Nicastro R. & C. s.r.l., founded in 1981, derives from another company working in the stone sector since 1976. The company handles and markets the production of slabs and finished pieces in marble and granite. Along with those cited, some of its major commissions were the pre-cast flooring for the Langham Hilton Hotel (the only six-star hotel in England), and the finished material for the Portman Ritz Carlton Hotel in Shanghai. International Italmarmi works with a quality system conforming to UNI EN ISO 9002 standards (1994) as officially certified by Sincert/Det Norske Italia s.r.l.

International Italmarmi di Nicastro R. & C. s.r.l.
Viale Roma 209 bis, 54100 Massa (MS), Italy
ph. +39 0585 254355; fax +39 0585 254307
Internet: http://www.internationalmarmi.com; e-mail: info@internationalmarmi.com

Calacata Sponda

ESTHETIC CHARACTERISTICS

Statuario has always been the white marble par excellence, a pure white stone without shadings or veins that could mar the faces of statues, and with a structure suitable for fine work with a chisel. For ornamental use the presence of small and discontinuous gray veins does not compromise its quality; if numerous, the veins form a pattern that stands out from the white ground, enlivening the material's design in an allover view. In this case the marble takes the name of Statuario Venato.

PRODUCTION

Statuario is extracted in various parts of the Carrara marble basin. The zone richest in this valuable marble is Monte Betogli, halfway between the Torano and the Miseglia basins. Total production of Statuario, including Statuario Venato, is about 4,000 to 6,000 tons a year. Blocks are medium to large, shapeless or squared.

WORKMANSHIP AND USES

Statuario is worked in tiles and medium- to large-sized slabs. It is recommended for facing in interiors and as low-traffic flooring. Obviously, it is widely used for statues, and in small blocks for fireplaces and special works in general. It is always polished or, at least, fine-smoothed.

HISTORY

Like many Carrara marbles, Statuario was already known to the Romans, at least from the first century B.C., and in some quarries you can still find traces of excavation from those and subsequent times. Among the main historical works we cite the Pantheon (the pronaos capitals), the Apollo Belvedere in the Vatican Museums, and Bernini's statue of St. Teresa in the Church of Santa Maria della Vittoria in Rome.

WORKS

In addition to works created in former times, Statuario has been used for a great many modern ones. It is especially popular in the Arab world where it has been used for numerous prestigious settings and in the private residences of sheiks and emirs. Among the most significant of these are the Arab Organization Headquarters Building in Kuwait (Statuario Venato), the Bait Al Bahjat Al Andhar in Oman, the Gracechurch City Office in London, and the Y.S. Club in Taipei (Statuario Venato).

THE COMPANY

Aldo Vanelli Marmi di Giorgio Vanelli is one of the most important producers of Statuario and can boast of a long tradition in extracting this material. The quarry, in operation for a great many years, has belonged to the Vanelli family since 1936 and also produces Venatino Betogli and Statuarietto. Output is handled exclusively by another family-owned company, Statuaria Marmi s.r.l., which markets it internationally as raw and finished product.

Aldo Vanelli Marmi di Giorgio Vanelli
Statuaria Marmi s.r.l.
Via Ilice 15, 54033 Nazzano Carrara (MS) Italy
ph. +39 0585 846300; fax +39 0585 846300
Internet: http://www.statuariamarmi.it; e-mail: info@statuariamarmi.it

Statuario

ESTHETIC CHARACTERISTICS

What distinguishes Rosa Aurora is the delicate color of its ground, which is a very pale pink tending at times to cream, uniform or at the most interrupted by barely visible shadings of white. The material is given vivacity and movement by tiny, very shaded, reddish streaks, and dots that tend to form wavy, slightly oriented veins. Grain is medium.

PRODUCTION

Rosa Aurora is one of the finest varieties of Rosa Portugal and is extracted in the basin lying between the towns of Estremoz and Vila Viçosa in the Alto Alentejo region of Portugal. Annual production is some thousands of tons in medium or small squares and shapeless blocks.

WORKMANSHIP AND USES

Rosa Aurora can be worked as slabs or tiles for use as low-traffic flooring and as facing. It is often worked as small blocks for tops and bathroom fixtures. It should be used solely in interiors.

WORKS

Rosa Aurora is used most frequently in luxury furnishings for private homes, such as the Storey Residence in the Philippines.

Rosa Aurora

ESTHETIC CHARACTERISTICS

Azzurro d'Oriente is formed of an alternation of azure and white veins, the former prevailing over the latter. This medium-large grain marble has a delicate look but also one of great character. In overall appearance, the color varies from pale sky blue to deep azure, sometimes mixed with white in straight or definitely undulating bands, but more often forming a slight pattern. This is accentuated here and there by brown streaks, the result of oxidation processes, which give the pattern movement and highlight the material's chromatic range. Grain runs from medium to large.

PRODUCTION

Although limited to only 25 m^3 a month, Azzurro d'Oriente's production guarantees a constant supply. Blocks are always well-squared and run from an average 220 x 130 x 100 cm to a maximum of 300 x 150 x 120 cm.

WORKMANSHIP AND USES

Azzurro d'Oriente is primarily worked in medium to large slabs, sizes that allow for the best appreciation of the material's appearance. Tiles can be made with it, but are less frequent and must always be cut *al verso* (with the grain). Both slabs and tiles are used indoors for facing and flooring. The surface is always polished.

THE COMPANY

Azzurro d'Oriente is handled exclusively by Block & Rock s.r.l., which markets it in raw block and slabs. Founded in 1993, Block & Rock immediately caught the international market's attention for its great organizational abilities and business dynamism. It makes special efforts in searching for new materials (of which Azzurro d'Oriente is a prime example) to offer its clientele a varied and up-to-date assortment of materials that meet with growing market demands.

Block & Rock s.r.l.
Via Antica Massa 38, 54031 Avenza, Carrara (MS), Italy
ph. +39 0585 52883 r.a.; fax +39 0585 856708;
Internet: http://www.blockrock.com; e-mail: blockroc@tin.it

Azzurro d'Oriente

ESTHETIC CHARACTERISTICS

Azul Macaubas is distinctive for its color: a pale azure-to-white ground with bright azure veining—veining that in places takes the form of streaks with vague outlines, in others as clear-cut threadlike lines. The veins run fairly straight or in wide curves, converging into one another, and always in a definite direction. This creates an oriented and slightly wavy pattern, accentuated in overall appearance by variations in the brightness of the color.

PRODUCTION

Azul Macaubas is extracted in a quarry near the town of Macaubas, in the state of Bahia, Brazil. The quarry's average yearly output is several thousands of tons, in blocks highly variable as to size and volume.

WORKMANSHIP AND USES

To enhance its esthetic features, Azul Macaubas is usually worked in slabs and, if used in customized jobs, pieces are sized to the dimensions of the azure veins. It is generally used for luxury flooring and facing and in bathroom fixtures, but often also in special works and premium statuary for outdoor use.

WORKS

Azul Macaubas has been used in a great many works throughout the world. One of the most significant is the cladding on the new head offices of Ambiente International Inc., in Tokyo, designed by Aldo Rossi and Cappa Kitai Architects and Planners.

Azul Macaubas

ESTHETIC CHARACTERISTICS

A lively, ceaseless succession of azure, lavender, green, and (to a lesser extent) white veins—in both thin and large bands, sometimes straight and sometimes in more or less accentuated curves. This, in essence, is Azul Imperial, a material of great esthetic effect and changing character. It is strong where the azure vein prevails and delicate where the different colors combine, shading into one another or forming threadlike, wavy lines. In any case the pattern is an oriented one and gives the material an outstanding feeling of movement. The grain is medium-fine.

PRODUCTION

Azul Imperial is extracted from a quarry in Bahia, Brazil. Monthly output is about 70 m^3 in squared blocks sized from an average 250 x 130 x 100 up to a maximum of 300 x 180 x 150 cm.

WORKMANSHIP AND USES

It is advisable to work Azul Imperial in big slabs cut either with the grain or the hard way. As slabs it is primarily destined for luxury interiors as flooring and facing, and is also frequently used in bathrooms where the material's alternating colors make for especially interesting esthetic effects. It is also commonly worked in small blocks for special interior pieces, and sometimes for quality manufactures for urban décor.

WORKS

Azul Imperial has been used in numerous private dwellings and in projects such as all the Volksbank branches in Nurnberg, Germany.

THE COMPANIES

Azul Imperial is marketed jointly and exclusively by the Pelé Granits Group and Rossittis. The Pelé Granits Group is one of Europe's major suppliers of granite, quartzite, and marble. In addition to Azul Imperial Pelé has the exclusive for Saga Blue Pearl, which it extracts on its own in Norway. The group has two sawmills in France and one in Italy run by its affiliate Pelé Granits Italia. This company specializes in furnishing international markets with 2- to 3-cm thick slabs of about fifty varieties of high quality granite, marble, and quartzite. It has over 2000 m^2 of Azul Imperial constantly on hand, which can also be viewed on the Internet.

Rossittis is one of Germany's largest dealers in natural stone. Its 35,000-m^2 warehouse, located near Holzwickede, has over 160 varieties of natural stone selected from the finest on the international market. The warehouse is constantly restocked and holds about 200,000 m^2 of slabs, 160,000 m^2 of tiles, and 80,000 m^2 of flooring.

scale 1:6

Pelé Granits Italia s.r.l.
Loc. Piano della Quercia 54016 Licciana Nardi (MS) Italy
ph. +39 0187 471784; fax +39 0187 471793
Internet: http://www.pelegranits.com; e-mail: info@pelegranits.com

Rossittis-Naturstein Import
Stehfenstrasse 59-61 59439 Holzwickede Deutschland
ph. +49 2301 91332-0; fax +49 2301 91332-32
Internet: http://www.rossittis.de; e-mail: throho@cityweb.de

Azul Imperial

ESTHETIC CHARACTERISTICS

Quarzite Blu is formed by a series of bands that run from white to deep blue, passing through an infinite number of intermediate shadings. The bands vary in thickness and at times are either nearly straight or undulating, depending on whether the block is cut along the hard way or the grain. Often, within an individual band, the blue is concentrated in thin veins that emphasize the movement.

PRODUCTION

Quarzite Blu comes from Madagascar. The quarry lies on the flanks of an enormous rocky mass in the central region of the country. Production is about 300 tons per month, 50 percent of which is prime quality. Blocks are generally an average 280 x 130 x 100 cm and are well-squared.

WORKMANSHIP AND USES

Quarzite Blu is commonly worked in slabs of different thickness and in commercially sized tiles. It is utilized primarily in flooring and facing, using layouts that highlight the material's special delicate color. Its appearance and physical properties also make it suitable for bathroom fixtures (tops and facing) and in luxury furnishings in the form of special elements, even in small blocks. Surface treatments are polished and smoothed; less frequently, a split finish is used. Treatment with an impregnating protective substance is often recommended.

WORKS

Since the material has only recently come on the market, there are no significant works yet to note. It is increasingly used in homes.

THE COMPANIES

Quarzite Blu is available in blocks and slabs in different thicknesses from Solmar S.p.A. and Calvasina S.p.A. The two companies worked for a long time in the marble industry, a past that has given them traditions and experience rarely equalled worldwide. The extraction division has three quarries: Quarzite Blu, Rosa Limbara granite in Sardinia, and Pietra Dorata sandstone in Tuscany.

On the processing level, Solmar and Calvasina specialize in raw, bushhammered, flamed, smoothed, and polished slabs of granite, marble, and stone, which they market worldwide. In this field Calvasina S.p.A. has a patent for 5-mm slabs turned out by a gangsaw and reinforced with carbon fiber or fiberglass.

Solmar S.p.A.
Via Milano 94, 22063 Cantù (CO) Italy
ph. +39 031 730373; fax +39 031 730056
e-mail: solmargraniti@tin.it

Calvasina S.p.A.
Via Promessi Sposi 10, 23868 Valmadrera (LC) Italy
ph. +39 0341 581124; fax +39 0341 200175
Internet: http://www.calvasina.com; e-mail: info@calvasina.com

Quarzite Blu

ESTHETIC CHARACTERISTICS

Rosso Carpazi is an unusual breccia, in which the elements and ground mass have practically the same color—a deep bright red. As a result, from a distance it appears to be a quite uniform marble with an almost austere appearance, given slight movement by a delicate weave of white veining. Looking closer, one can admire the brecciated pattern better, and with it the elegance of the shadings and contrasts of color that only a breccia can offer.

PRODUCTION

Rosso Carpazi is extracted in Albania, near the town of Muhur in the Korabi Peskopje region. The quarry produces about 4,000 tons a year in squared blocks sized a maximum 290 x 130 x 120 cm.

WORKMANSHIP AND USES

Rosso Carpazi can be worked as slabs and tiles without compromising its esthetic value. In both cases the material must be reinforced with resin and webbing and can be used for flooring and facing, paving and cladding. It is also frequently worked as solid stone for columns, tables, vanity tops, and special jobs in general, to decorate interiors and bathrooms.

WORKS

Among the main works using Rosso Carpazi are the Parliament Building in Berlin designed by T. Van Valentine, Marconi Airport in Bologna designed by L. Sacchetti, and the I Gigli shopping center in Florence, designed by A. Natalini.

THE COMPANY

Rosso Carpazi is extracted by UpMuhur srl, a member of the UpGroup s.r.l. The latter includes several companies present on the market for over thirty years, with businesses that run from extracting marble in the Apuan basin to producing and installing flooring and facing to creating art objects in marble and stone. Among the most important works the group has contributed to are the ICE (Italian Foreign Trade Institute) offices in New York City; the Beethoven Hause Museum in Bonn; the Münich, Cologne and Bonn airports; the Florence Gift Mart offices in Florence; and Linate and Fiumicino airports in Italy.

UpGroup s.r.l.
Via Acquale3, 54100 Massa Italy
ph. +39 0585 831132; fax +39 0585 832038
Internt: http://www.upgroup.it; e-mail: info@upgroup.it

Rosso Carpazi

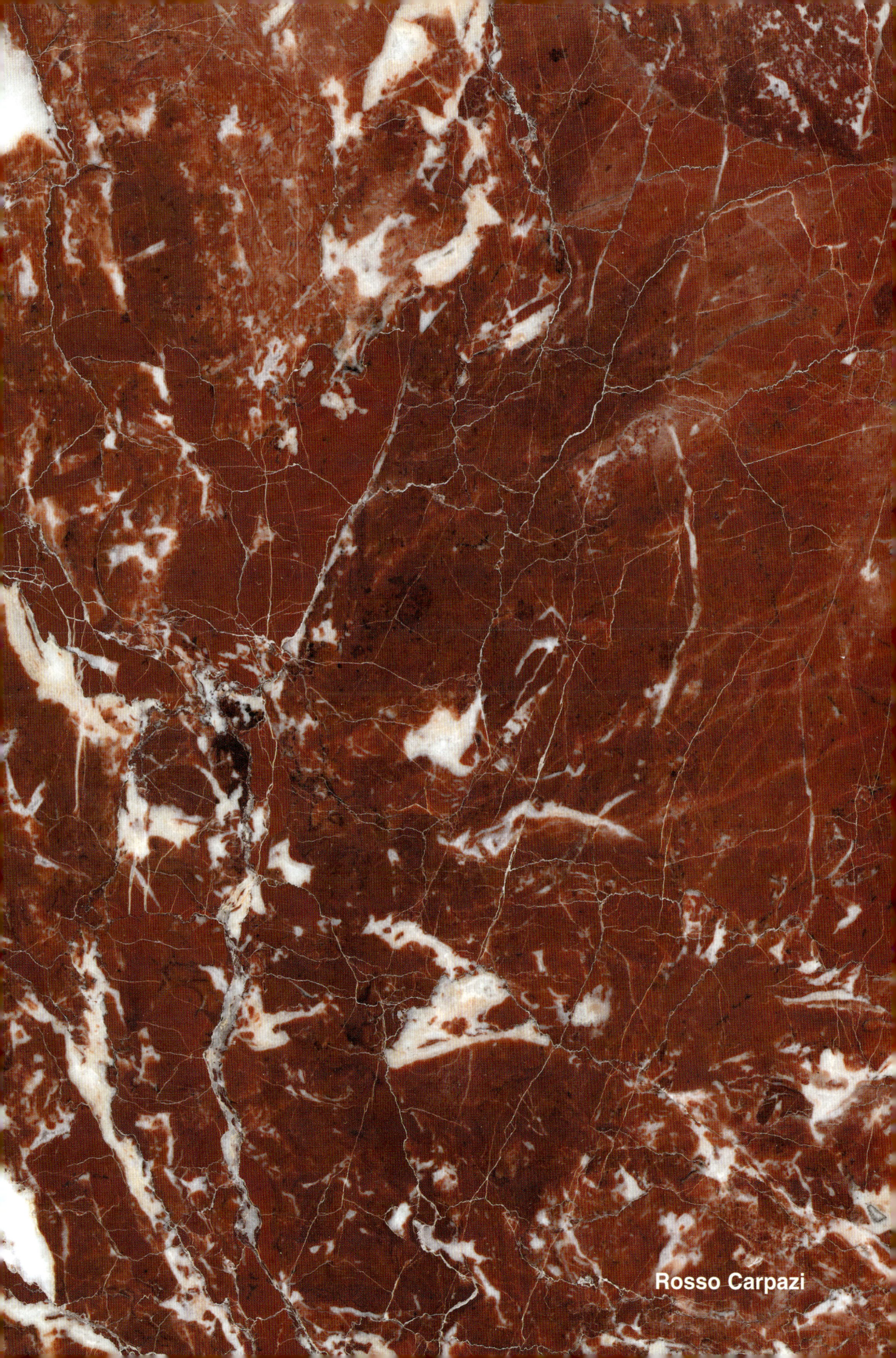

ESTHETIC CHARACTERISTICS

Rouge France Incarnat is a marble with a strong personality, imposing with a deep yet bright brick red ground. On it are highlighted shapeless, elongated white elements often placed to give direction to the material's pattern. White veins cross the ground every so often in a wavy movement, and tiny white dots on the ground lighten up the overall chromatic effect.

PRODUCTION

Rouge France Incarnat is extracted near Saint-Nazare de Ladarez, a town near Béziers in the south of France. The output of the only working quarry is about 150 tons a month in blocks sized up to 280 x 150 x 80 cm.

WORKMANSHIP AND USES

Rouge France Incarnat can be worked in either slabs or tiles for use as flooring and facing, often in combination with other materials. It is very frequently worked in small blocks for special decorative elements such as columns, fireplace coverings, vanity tops, and bathroom décor. It can also be used for urban fixtures. Outdoor as well an indoor applications are possible, although this particularly decorative material is usually found in interiors.

WORKS

Among the major works created with Rouge France Incarnat are the Louvre Museum in Paris; the West Hammersmith Centre in London; Place de la Loge in Perpignan, France; the Promenade Shopping Center in Toronto; and the Fréjorgues Airport in Montpellier.

Rouge France Incarnat

ESTHETIC CHARACTERISTICS

Rosso Collemandina has a homogeneous brick red ground with slight variations in color depth, on which a sparse weave of wide veins design an irregular layout similar to an opus sectile pavement. But more than veins, these are bands of a red paler than the ground but with a central core, dark red or white, giving the whole pattern body and luminosity.

PRODUCTION

Rosso Collemandina is extracted from a single quarry on the Apennine side of the Garfagnana, near the town of Villa Collemandina (Lucca). The quarry, partly worked underground, has an annual output of 2,000 tons. The colors of the other varieties tend towards brown and gray.

WORKMANSHIP AND USES

Rosso Collemandina is worked in slabs and tiles for flooring and facing. It is frequently made to measure for use in combination with other materials and worked in small blocks for special pieces such as the coverings of large columns. In any case the material is destined for interiors. The surface can be polished, smoothed, or antiqued.

WORKS

Among the most important works using Rosso Collemandina are the Park Hyatt Hotel in Melbourne, Australia and 101 West End Avenue in New York City.

THE COMPANY

Rosso Collemandina is extracted exclusively by Gino Corsi e C. s.a.s. Founded in 1966, the company specializes in standard and made-to-measure colored marble products. It also has a Pietra del Cardoso quarry and markets marbles in blocks and slabs. In addition to the works just cited, the company supplied materials for the casino in Melbourne, Australia; the Intercontinental Hotel in Houston, Texas; the courthouse in Arlington, Virginia; and the Feltrinelli Bookshop in Milan.

Gino Corsi e C. s.a.s.
Via San Bartolomeno 32, 55045 Pietrasanta (LU) Italy
ph. +39 0584 71180/71674; fax +39 0584 72253

Rosso Collemandina

ESTHETIC CHARACTERISTICS

Rosso Verona is the head of an esthetic family composed of many ornamental materials, but it is unequaled for the delicacy and elegance of its patterns and colors. On a warm, brick-red ground with pastel tones appear rounded, sometimes lobed, shapes with clear-cut outlines, in a paler red than the surrounding mass. Depending on cutting direction, the shapes are arranged in long bands or an irregular grid. The result is a sober, serene appearance, unexaggerated but very attractive. Grain is fine.

PRODUCTION

Rosso Verona is extracted from the hills surrounding S. Ambrogio di Valpolicella in the province of Verona. Production, along the lines of 10,000 tons a year, can seesaw quite a bit as deposit conditions change. Blocks often have a low side but at least two others of sufficient size.

WORKMANSHIP AND USES

Rosso Verona can be worked into tiles or medium-sized slabs. Quite frequently it is made to measure and combined with other materials and used in small blocks for tops, columns, fireplaces, and other special works. Surface treatments can be polishing, smoothing, or shock. The material is often *magroso* and generally needs to be stuccoed, resin-coated, and reinforced with webbing. It can be used for flooring and facing.

HISTORY

Rosso Verona was already known in Roman times as a marble that current historians define as a substitute. In fact, its exploitation began in the Imperial period not so much at Rome's behest as to make up for the lack of colored marbles being supplied to northern Italy by the Imperial quarries, mainly located in the eastern provinces.

WORKS

Among the innumerable works created with Rosso Verona are obviously many important buildings in Verona's historical center. The most important recent works include Four Millbank in London, the Church of San Domenico Savio in Verona, the Verona Chamber of Commerce, and the Al Ishaa Guest Palace Hotel in Saudi Arabia.

Rosso Verona

ESTHETIC CHARACTERISTICS

Although formed of shapeless large and small lithoid elements, abutting one another in no defined order, Breccia Pernice is quite constant in its esthetic model, which in an overall view has essentially uniform design. The color of the classic variety (Breccia Pernice Scura) is a blend of the more or less deep red of the mass and some elements, and the bright pink of the larger elements.

In the lighter variety (Breccia Pernice Media), the pink component predominates, both in the elements and in the mass. In both cases the material is outstanding for its warm, relaxing colors and the elegance of the brecciated pattern.

PRODUCTION

Breccia Pernice is extracted on the slopes of Monte Pastello in the community of Fumane (Verona). Of the three quarries currently operating, only one produces the classic, dark red type. Blocks are both shapeless and squared and sized up to 300 x 180 x 180 cm.

WORKMANSHIP AND USES

Breccia Pernice is worked in slabs and tiles. In these forms it is mainly used for flooring and facing; as slabs, it can be used to good effect in mirror-image applications. In any case the material has to be resin-coated and reinforced with webbing. It is frequently worked in small blocks for fireplaces, table and bathroom tops, and articles of religious art.

HISTORY

Breccia Pernice has been extracted from its current site since the late 1900s and reached its peak in the mid-twentieth century.

WORKS

Among the many works created with Breccia Pernice are the Church of San Giovanni Rotondo in Foggia, Italy (monolithic columns and the altar facing); Trump Tower in New York City; the James Center in Richmond, Virginia; Piccadilly Center in Sydney, Australia; the Four Seasons Hotel in Berlin; and the Ritz Carlton Hotel in Osaka.

THE COMPANY

Breccia Pernice is extracted exclusively by EssegiMarmi di Giulio Savoia e C. With its affiliate, Sottilmarmo Savoia, the company specializes in working premium and colored marbles, including Rosso Verona and Verde San Nicolaus. It works full cycle, from extraction to the finished product (also made to measure), including special dimension work. On the commercial level, the company offers both raw blocks and semi-finished and finished pieces. In addition to the aforementioned works, the company also contributed to creating many other prestige projects such as the Ritz Carlton hotels in New Orleans, Aspen, and Jamaica; the Monte Carlo Hotel in Las Vegas; and the Hotels Europa and Regina in Venice.

EssegiMarmi di Giulio Savioa e C. s.a.s.
Via Passo Napoleone 521, 37020 Volargne (VR) Italy
ph. +39 045 6860800 r.a.; fax +39 045 7732972
Internet: http://www.essegimarmi.it; e-mail: essegimarmi@iol.it

Breccia Pernice

ESTHETIC CHARACTERISTICS

Diaspro Tenero di Sicilia (Soft Sicilian Jasper) has an appearance quite similar in pattern and color to that of real jasper but is found here in a calcareous material ascribable to the family of marbles. It is a polychrome ensemble of great impact, in which especially bright red and yellow prevail over pinkish white, arranged in single scattered elements and in irregular veins but also in small bodies with onyx's typical design. Often, between the plays of color, there is a clear brecciated structure; in other places the red or yellow appears to dominate, forming a ground at times interspersed with spots of another color.

PRODUCTION

Diaspro Tenero di Sicilia is a very fine variety of the "Marbles of Custonaci," a group of marbles extracted in several quarries in the province of Trapani. Since the material is found in a thin layer between one marble and another, production may be discontinuous but amounts to 50 to 100 tons a year. Blocks are small to medium size, with lengths of 120 to 150 cm and heights of 60 to 110 cm.

WORKMANSHIP AND USES

The format best suited to Diaspro Tenero di Sicilia is of large polished slabs to be used in interiors. Mirror-image installation is most effective. It is also frequently used in small blocks for special works, again always for interiors. Part of production is destined for sculpture and use with antique furniture. The material has to be coated with resin and reinforced with webbing.

HISTORY

It is possible, but not certain, that Diaspro Tenero di Sicilia was already known to the Romans and included among the "Tauromenitana Marbles," a term denoting marbles from Sicily. What is sure is that the stone was popular from the Renaissance to the eighteenth century, as shown by the popularity of this material in many Roman churches. Examples include the church of St. Catherine of Siena in Monte Magnanapoli and the church of San Luigi de' Francesi.

WORKS

In addition to the historical works just mentioned, Diaspro Tenero di Sicilia has been used in many recent works running from embassies to banks to luxury hotels. It has also been used to decorate luxurious ships and private planes.

THE COMPANY

Diaspro Tenero di Sicilia is distributed exclusively by Antolini Luigi & C. S.p.A. This company, founded in 1956, specializes in supplying marble, granite, and stone in general as semi-finished and finished products and, through its subsidiary, Eurotrading S.p.A., as blocks. The company normally offers over 450 varieties of materials, many exclusively worldwide, and owns a Rosa Beta quarry and another of Serizzo Antigorio. Its four processing plants have a total of twenty-seven gangsaws and six slab lines combined with many machines for processing and special treatments.

Antolini Luigi & C. S.p.A.
Via Marconi 101, 37010 Sega di Cavaion (VR) Italy
ph. +39 045 6836611; fax +39 045 6836745
Internet: http://www.antolini.it; e-mail: marketing@antolini.it

Diaspro Tenero di Sicilia

ESTHETIC CHARACTERISTICS

A prime feature of Brèche de Vendôme (also called Brèche de Benou) is undoubtedly its extraordinary polychromy, often associated with an uneven distribution of colors and forms. In a short stretch bands of shapeless and fringed violet elements immersed in a greenish mass change to zones of a compact-looking ivory crossed by golden-yellow streaks. In places, roundish black shapes run in sets, accentuating the material's directional pattern. In brief, it is an extremely variegated chromatic and structural whole but one able to maintain the constancy required for commercial use.

PRODUCTION

Brèche de Vendôme is extracted in a quarry south of Pau in the Pyrenees region of France. Currently, the annual output is about 300 to 400 m^3. The blocks are often well-squared and vary in size from 200 to 300 x 100 to 106 x 120 cm.

WORKMANSHIP AND USES

To enhance its esthetic characteristics to the utmost, Brèche de Vendôme is mainly worked in large slabs and given a polished surface. In this form it is used in interiors requiring vivacious colors but also a warm ambience, perhaps tending to the classic. In fact, it is quite an appropriate material for furniture tops in Louis XIV, Louis XV, and Louis XVI style. It is also often used for luxury facing and flooring and is quite effective when installed in mirror-image layouts.

HISTORY

Brèche de Vendôme was already exploited at the time of Henri IV (1200–1300) and was widely used at Versailles and in the Royal Palace of Laaken in Belgium.

WORKS

In addition to the historical works just cited, Brèche de Vendôme was used to face some rooms in the White House in Washington D.C. and to adorn many of the most beautiful buildings and churches in Paris.

THE COMPANY

Brèche de Vendôme is distributed exclusively by Antolini Luigi & C. S.p.A. This company, founded in 1956, specializes in supplying marble, granite, and stone as semi-finished and finished products and, through its subsidiary, Eurotrading S.p.A., as blocks. The company normally offers over 450 varieties of materials, many exclusively, and owns a Rosa Beta quarry and another of Serizzo Antigorio. Its four processing plants have a total of twenty-seven gangsaws and six slab lines combined with many machines for processing and special treatments.

Antolini Luigi & C. S.p.A.
Via Marconi 101, 37010 Sega di Cavaion (VR) Italy
ph. +39 045 6836611; fax +39 045 6836745
Internet: http://www.antolini.it; e-mail: marketing@antolini.it

Brèche de Vendôme

ESTHETIC CHARACTERISTICS

Although its structure is quite similar to normal beige travertine, Travertino Rosso differs in color. The red is not uniform but varies in intensity and tone, especially at right angles to the material's different layers, whose orientation is highly accentuated by the layout of the cavities typical of these materials. In places the red color takes on an orange tone and, more rarely, passes decidedly into yellowish streaks.

PRODUCTION

Travertino Rosso is extracted in Iran, near the city of Tabriz. Average production is between 6,000 and 10,000 tons a year in medium and small blocks, generally well-squared.

WORKMANSHIP AND USES

Travertino Rosso is commonly worked in both slabs and tiles. The material can be polished perfectly and is often smoothed. It is chiefly destined for flooring and facing, excluding bathrooms and kitchens, but is also frequently used for tables, tops, and special works in general. Its visible side usually needs to be stuccoed.

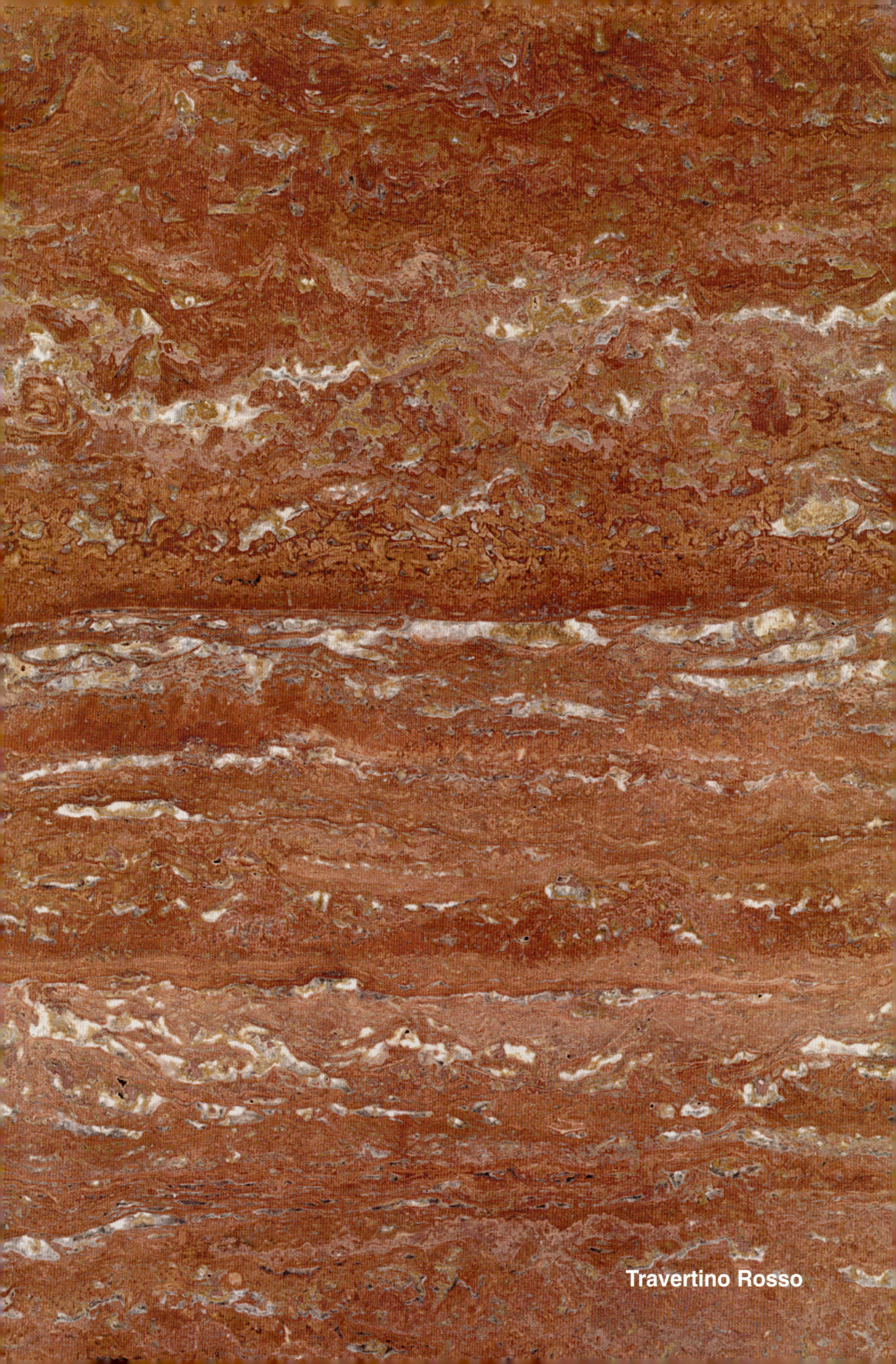

Travertino Rosso

ESTHETIC CHARACTERISTICS

Fior di Pesco Classico Apuano is a very special type of breccia originating from the thickening of often roundish elements but also of jagged masses of no definite shape. Their color, a lovely purplish red of a constancy unusual for a breccia, shades (or rather, passes) into the statuary white of the ground, in the shape of delicate flakes imbedded between one element and another. In places the overall chromatic effect is enriched by greenish veining that winds over the purplish mass in thin lines or in large, more or less undulating areas.

PRODUCTION

Fior di Pesco Classico Apuano is extracted from the same quarry that produces Breccia Capraia (see entry for specific information). Fior di Pesco production is about 600 to 800 tons a year of squared blocks that are not very thick (about 80 cm) but about 2 meters wide and even several meters long.

WORKMANSHIP AND USES

Worked in slabs, Fior di Pesco Classico Apuano is primarily used as facing and flooring in luxury interiors. Sawed along the grain it provides large slabs that are quite homogeneous in appearance. Sawed along the hard way, the material forms a more or less thick band bordered by the statuary white ground; in this case its mirror-image installation is very effective. It is also popular in special works for rooms decorated with antiques, such as fireplaces, columns, and tops for period furniture.

HISTORY

Fior di Pesco Classico Apuano is the only member still on the market of a family of materials extracted in several Apuan areas since the Renaissance, of which Breccia Medicea is perhaps the best known. Its name, but also its esthetic features, are reminiscent of the famous Marmor Chalcidicum (also known, like this marble, as Fior di Pesco) extracted in Greece. It was one of the finest marbles known in Roman times and was reused in the Baroque period to decorate several churches in Rome.

WORKS

Fior di Pesco Classico Apuano has always been used to decorate churches and historic buildings as well as private homes worldwide. It has also recently been utilized in the furnishings of luxury yachts.

THE COMPANY

Fior di Pesco Classico Apuano is extracted and sold by E.B.C. of Massa. The company, founded in the early 1970s, is part of the GMC Group which includes three other concerns with premises in, respectively, Ortonovo, Massa, and Porto Torres. The group also owns a Bianco Carrara quarry and two of Sardinian granite (Rosa Beta and Rosa Antico). In its own plants the group goes through the entire production cycle, from extraction to the finished product, working both marble and granite. Among its major commissions were the New China Building in Guangzhou, China; the Taiwan Cement Building in Taipei; the new Communist Party headquarters in Shanghai; and the Pulawska Financial Center in Warsaw.

GMC di Grassi Luciano & C. S.p.A.
Via Fossone Basso 8, 19034 Isola di Ortonovo (SP) Italy
ph. +39 0187 661680; fax: +39 0187 66350
e-mail: info@gmcspa.com

Fior di Pesco Classico Apuano

ESTHETIC CHARACTERISTICS

Salomè forms a close-knit pattern of streaks or actual veining on an elegant gray ground. The veins are often reddish but also whitish and tend in places to a yellow color. The pattern looks uneven, jagged, and takes on an overall brecciated appearance, given specific movement by the direction most of the veining takes.

PRODUCTION

Salomè is extracted from several quarries in Eskesehir in Anatolia, Turkey. Total annual production is around 10,000 tons of medium-sized blocks.

WORKMANSHIP AND USES

Salomè can be worked as either slabs or tiles without jeopardizing its esthetic value. Less frequently, it is worked as solid stone, mainly for vanity tops and small columns for special interiors. Polishing enhances the appearance of the material, which is used mainly for interior decoration.

WORKS

Salomè has been used to decorate the interiors of luxury homes. An important public work that utilizes Salomè is the Middlesex Building in Hartford, Connecticut.

Salome

ESTHETIC CHARACTERISTICS

Giallo Reale is distinctive for its pastel yellow ground whose chromatic intensity differs in zones that mix together in no apparent order—sometimes in big shapeless blots, other times in more defined and circumscribed elements. In places, and with variable frequency, the yellow shades into a fairly deep pink that enlivens the overall chromatic picture. An absence of shadings gives us the subvariety, Giallo Reale Golden.

PRODUCTION

Giallo Reale is extracted from a few quarries on the slopes of Monte Lessini, in the province of Verona. Annual production is some 10,000 tons, in blocks that run from shapeless to squared and quite large.

WORKMANSHIP AND USES

Giallo Reale can be worked as both slabs and tiles for use as flooring and facing. It is very frequently worked in thick pieces for fireplaces, kitchen counters and vanity tops, and split for cladding tiles (spaccatello). The material is usually resin-coated and given webbed backing only if worked as large-sized slabs.

WORKS

The most important recent work using Giallo Reale was the restoration of the cathedral in Caserta, Italy.

THE COMPANY

Giallo Reale is extracted by Eurocave s.r.l. by authorization of Fincave s.r.l., which holds the quarrying concession.

Eurocave s.r.l.
Viale del Lavoro 36, 37030 Colognola ai Colli (VR) Italy
ph. +39 045 6152264; fax: +39 045 6171499

Fincave s.r.l.
Viale Crosara 35, 37020 Luco di Grezzana (VR) Italy
ph. +39 045 8802091; fax: +39 045 8819175

Giallo Reale

ESTHETIC CHARACTERISTICS

Pietra Dorata is a sandstone with veinings that overlap in curvilinear bands, running from golden yellow to brownish and varying in thickness from several millimeters to a decimeter. The bands occur here and there, giving the material a delicately undulating look. On the upper edge of each band is a dark brown border, often several millimeters thick, which accentuates the undulating pattern. The color and pattern of the material are quite similar to those of wood, arousing the same sensation of warmth and comfort.

PRODUCTION

Pietra Dorata is extracted in Tuscany, particularly in Manciano in the province of Grosseto. The quarry, active since 1965, turns out up to 600 tons a month. About 50 percent of production consists of prime-quality blocks, squared on the average into sizes of 260 x 120 x 100 cm. Semi-squared blocks account for 30 percent; the remaining 20 percent are in shapeless ones. The same area also contains Pietra Etrusca and Pietra Maremma (Maremma Stone), other sandstones with different coloration.

WORKMANSHIP AND USES

Pietra Dorata is worked in slabs from two to ten cm thick, producing tiles in various formats. These tiles are primarily used in internal and external coverings and for low-traffic flooring. It is also frequently worked in small blocks to make shelves, table tops, columns, monuments, and other general objects. Special works include elements for fireplaces. The surface is usually smoothed, to a greater or lesser extent; less frequently it is bushhammered or split. Treating it with a protective impregnating substance is recommended.

WORKS

Among the most recent and significant works created with Pietra Dorata are the flooring (12,000 m^2) of the Malpensa 2000 airport in Milan (Studio Sottsass, Milan) and in the Tel Aviv synagogue designed by M. Botta. Other important uses have been as cladding on public and private buildings in Italy and elsewhere—for example, on the Banca d'America e d'Italia in Prato (Florence); the Cariplo Bank in Mariano Comense (Como); the Stadtsparkasse Bank in Wuppertal; the Sparkassenverwaltunsbau Bank in Munster; many buildings in Milan in Via Vittor Pisani and Via Molino delle Arme; and the interiors of churches, restaurants, shops, etc.

THE COMPANY

Pietra Dorata is extracted and sold in blocks of various sizes and types by the PIETRA DORATA company of Manciano (Grosseto).

Pietra Dorata s.r.l.
s.p. Manciano-Farnese Km. 8, Manciano (GR) Italy
ph. +39 0564 629098; fax: +39 0564 629385

Pietra Dorata

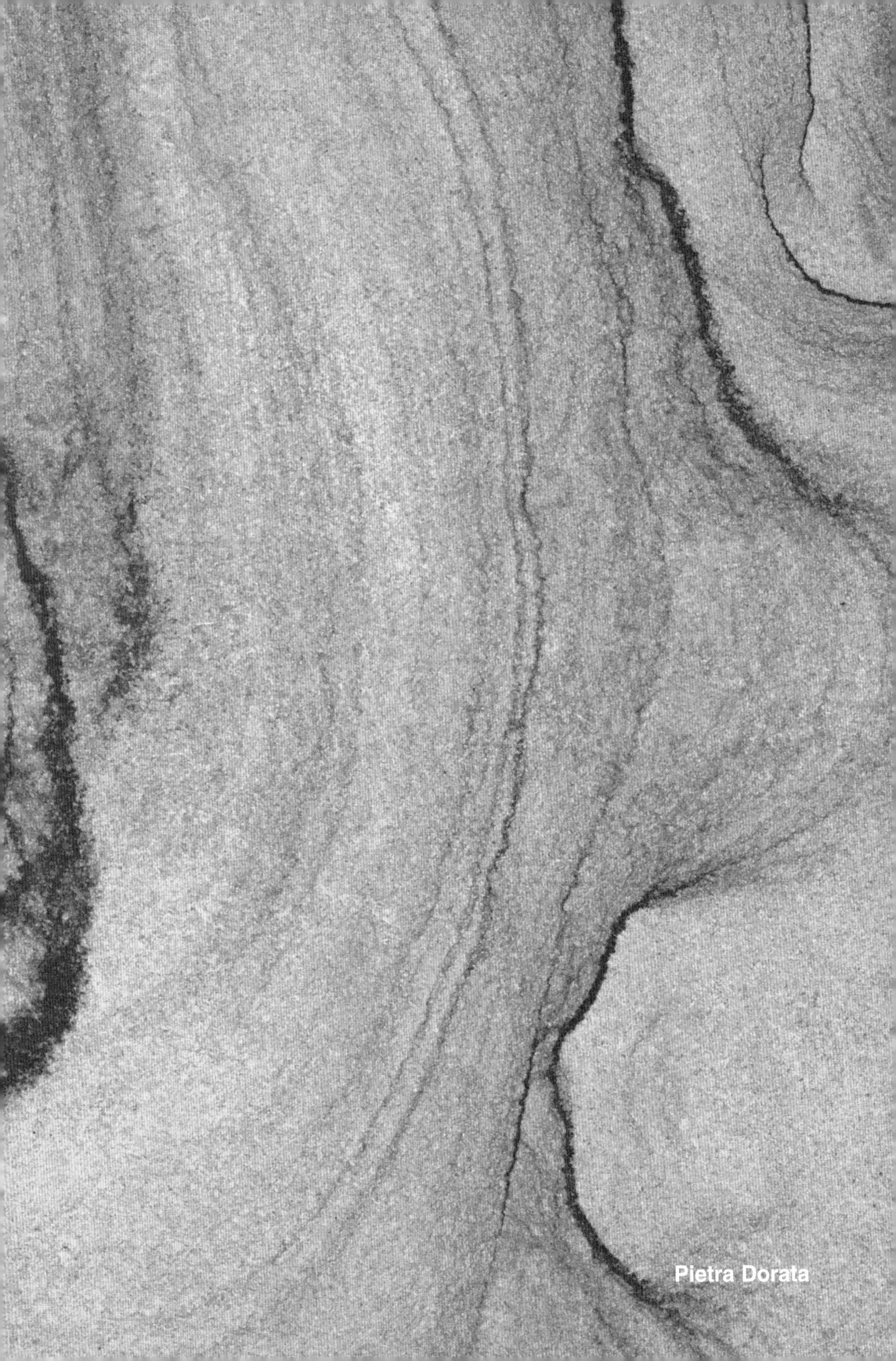

ESTHETIC CHARACTERISTICS

Travertino Dorato has a pattern similar to classic travertine but its various levels are a veritable kaleidoscope of different shades and depths of yellow. They go from pale or bright to dark yellow, and in some cases there are brownish tones. Its overall appearance is a golden yellow, bright but not too bright, luminous and at the same time restful. Cut with the grain, instead of overlapping, the colors appear in nebulous, uneven shadings.

PRODUCTION

Travertino Dorato is extracted in a quarry in Peru. Output is about 1,000 m^3 a year, in mainly medium-sized blocks.

WORKMANSHIP AND USES

Travertino Dorato is worked in slabs and tiles mainly for flooring and facing. It is frequently made to measure for special works in combination with materials of other colors. The material may need to be stuccoed.

WORKS

Two important works that use Travertino Dorato are the Bucherer jewelry store in Locarno, Switzerland, and the Banco Central de España in Madrid.

Travertino Dorato

ESTHETIC CHARACTERISTICS

What makes Giallo Siena a truly unique material is its coloring, which contains many tones and intensities of yellow. On a ground that goes from a pale, luminous yellow (sometimes with hints of ivory) to heavier tones that are almost brownish, bright yellow bands branch out, whose edges shade into the ground often without apparent interruption. This leads to a paradoxical situation in which, if looked at closely, the material looks variegated, while from a certain distance it appears to be monochromatic. Grain is very fine. Subvarieties have perceptible variations in color and perhaps some grayish veining.

PRODUCTION

Giallo Siena is extracted from a few quarries located on the hills of the Montagnola Senese, in the province of Siena. On the whole, production is often discontinuous and limited to several thousands of tons a year. Especially when the material is very bright in color, the blocks are often shapeless and small; squared blocks and those of medium size are more frequent in the paler varieties.

WORKMANSHIP AND USES

Due to its color and limited amounts, Giallo Siena is usually utilized as a decorative element in customized work, often together with other materials. It is frequently used as cut-to-size tiles for polychrome flooring and as solid stone for vanity tops and little columns. It is less frequently used in small slabs. In any case, Giallo Siena is suitable for flooring and facing but outdoor use is inadvisable.

HISTORY

Extracting yellow marble from the Montagnola Senese dates back to the Middle Ages, and it was developed mainly during the Renaissance, when Cosimo de' Medici I spurred a search for new colored marbles. Since then, the many varieties that gradually came to light have been given important uses, especially in decorating churches and basilicas in Rome and Florence. One example is the basilica of San Giovanni Bosco in Rome.

WORKS

In addition to adorning houses of worship, Giallo Siena has been used in its different subvarieties extensively in luxury interiors, both private and public. Some of its most representative uses were in the Langham Hilton Hotel in London, the Royal Holiday Inn and V.I.P. Club in Singapore, and the Colony Park Hotel in Geneva.

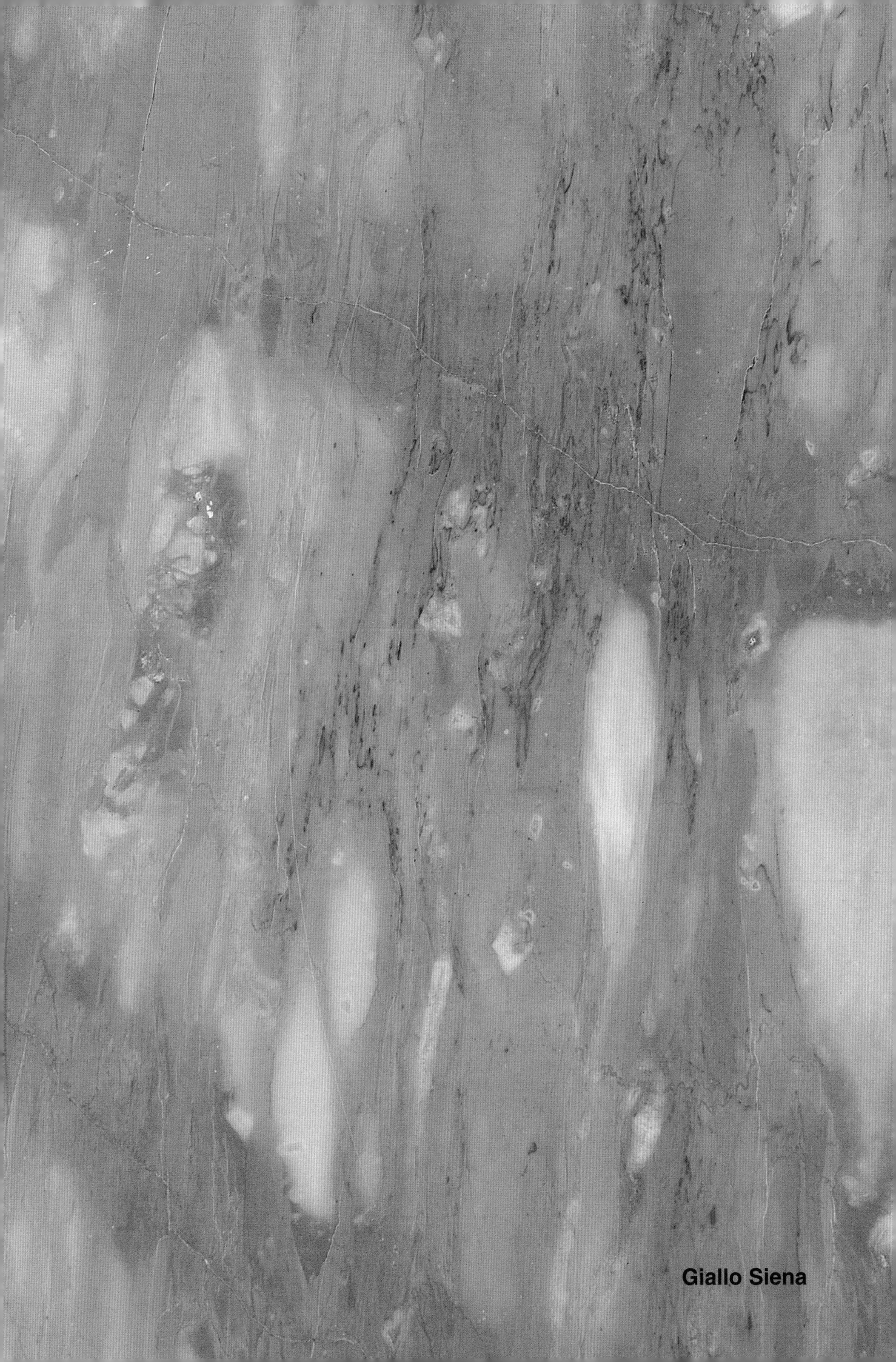

Giallo Siena

ESTHETIC CHARACTERISTICS

Cipollino Apuano is a vibrant material which, while following a well-defined esthetic model, never repeats itself, giving each piece unique patterns. It has a more or less deep olive green ground covered with thin, dark stripes running parallel to one another in sinuous, sometimes chaotic, runs but always in a prevalent direction. In places the stripes become so dense as to have the consistency of gray-green veins. White veining follows and accentuates the winding pattern of the ground and gives the material depth. But more than veins these are actually splashes of white, in some places taking the shape of twisted masses folding into themselves, and in others as thin fringing mixed with the green ground.

PRODUCTION

Cipollino Apuano is extracted in the Apuan Alps in Tuscany, Italy. It comes from a single quarry called the Pendia Tana located in the Turrite Secca valley on the Lucca side of the Apuan chain. In normal working conditions the quarry can produce up to 500 tons a month, 70 percent of which is prime quality. Squared blocks account for 85 percent of production and are sized 250 to 330 x 160 x 150 cm. The remaining 15 percent consists of semi-squared blocks.

WORKMANSHIP AND USES

The finished products are large slabs (up to and sometimes over 160 x 300 cm) cut both along the hard way and along the grain. In the first case, the pattern looks more chaotic and variations in the green tones of the ground are accentuated. In the second, the white veins look like better defined morphological and chromatic elements, emphasizing the material's overall orientation. Slabs are generally polished and used for flooring and facing, where they lend themselves to interesting mirror-image applications. If cut the hard way (frontally) the slabs have to be reinforced with webbing on their backs and, especially when there are *macchie lente* (loose blots), resin-coated on the surface to be polished. Even special thick pieces can be produced with this material, such as columns, tabletops, counters, and small blocks in various types produced for funerary uses. In these cases, too, the surface is usually polished, but interesting results can be achieved with shock treatments followed by a rinse of hydrochloric acid: This treatment highlights the white component rather than the green, generating a rough surface but keeping the color tones bright. A possible but infrequent use is in bathroom décor.

WORKS

Among the most prestigious works, it has been used to cover the walls and floors of the interiors of the Farnesina Palace (the site of Italy's Foreign Affairs Ministry).

THE COMPANY

Cipollino Apuano is produced exclusively by Cecconi Pietro & C. snc. The company, started in the early 1950s, currently produces blocks of Cipollino Apuano and Cipollino Cremo Tirreno as well as Bianco Carrara flooring tiles and skirting. In addition to supplying materials for the Farnesina, it also supplied the Cipollino Cremo Tirreno that decorates the press room in Palazzo Chigi in Rome, and the Cipollino Classico used to cover the ninety-five columns on the Australian Parliament building in Sydney.

Cecconi Pietro & C. snc
Via Lungofiume Versilia 15, 55045 Ponte Rosso di Pietrasanta (LU), Italy
ph. +39 0584 742649; fax +39 0584 742592

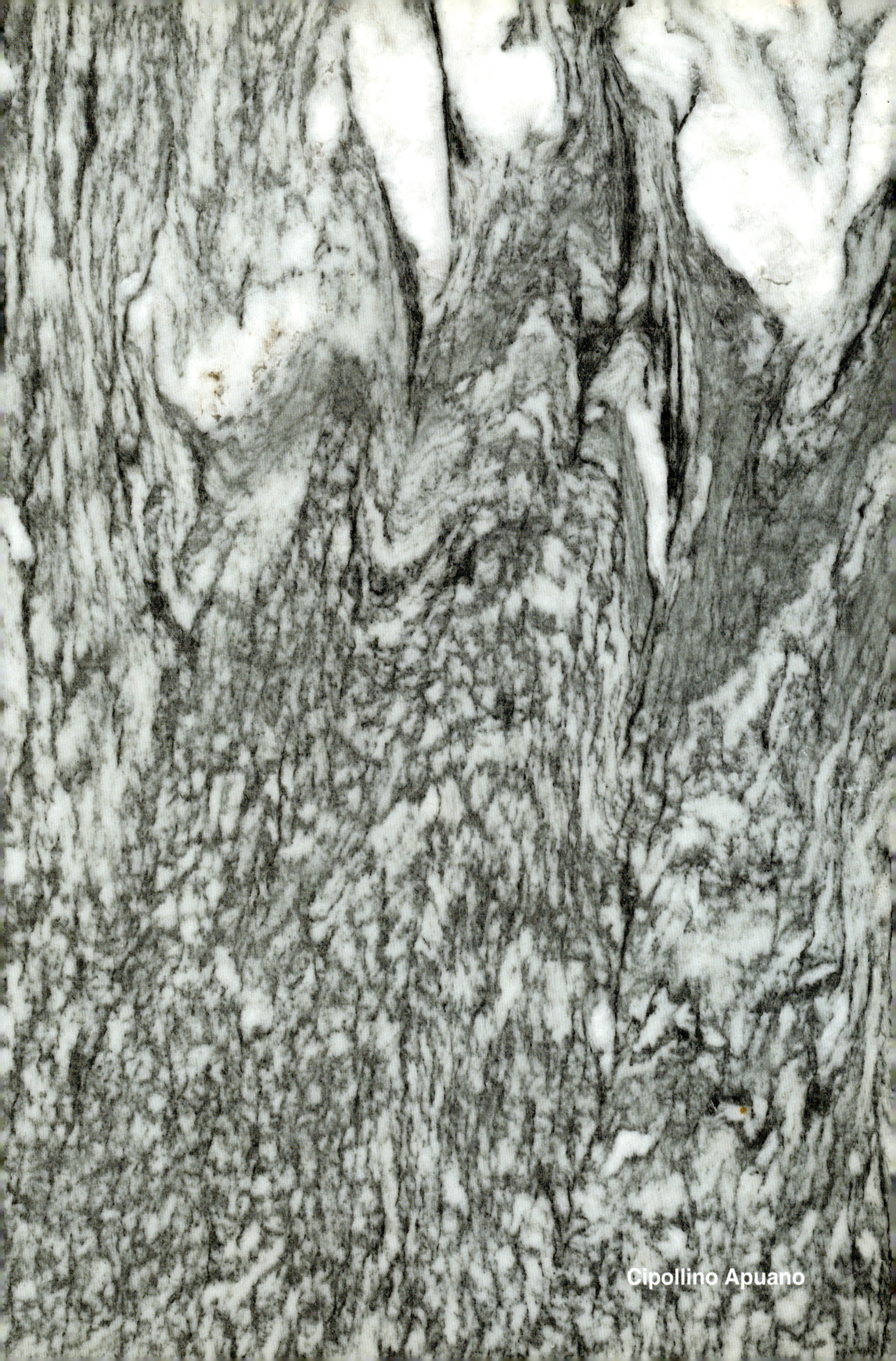

Cipollino Apuano

ESTHETIC CHARACTERISTICS

Cipollino Cremo Tirreno is characterized by a warm ivory ground mass with delicate greenish shadings, from which dark green veins stand out in quite variable shapes and thickness. These have a markedly undulating, at times twisting, run, forming a pattern like the sea in a storm. Not infrequently there are small thick and twisted veins of milk white chalcedony, among the dark veins that emphasize the whole material's movement.

PRODUCTION

Cipollino Cremo Tirreno is extracted in the Apuan Alps in Tuscany, Italy. It comes from a single quarry, called Gufonaglia, located on the Canale delle Fredde on the Lucca side of the Apuan chain. In normal working conditions production can reach 500 tons a month, 70 percent of which is prime quality. All of its output is in squared blocks sized 300 to 330 x 120 to 160 x 200 cm.

WORKMANSHIP AND USES

The finished products are generally large polished slabs (up to and sometimes over 160 x 300 cm), always cut along the hard way. In this form Cipollino Cremo Tirreno is used in flooring and facing where it is frequently installed in mirror-image pieces to great esthetic effect. The slabs usually require no consolidation or reinforcement treatments. Thick pieces are also produced for columns and tabletops.

HISTORY

Cipollino Cremo Tirreno has been extracted for the past twenty years or so. Although of relatively recent origin, its name and esthetic features can be traced back to an equally prized material, known and used in Roman times—Marmor caristium, extracted near Karystos, Greece, and also called cipollin for its particular, layered conformation similar to that of an onion.

WORKS

A significant work that utilizes Cipollino Cremo Tirreno is the facing on the pressroom in Palazzo Chigi in Rome.

THE COMPANY

Cipollino Cremo Tirreno is produced exclusively by Cecconi Pietro & C. snc. The company, started in the early 1950s, currently produces blocks of Cipollino Cremo Tirreno as well as Bianco Carrara flooring tiles and skirting. In addition to supplying materials for Palazzo Chigi, it also supplied the Cipollino Apuano that decorates interiors of the Farnesina Palace in Rome, and the Cipollino Classico used to cover the ninety-five columns on the Australian Parliament building in Sydney.

Cecconi Pietro & C. snc
Via Lungofiume Versilia 15, 55045 Ponte Rosso di Pietrasanta (LU), Italy
ph. +39 0584 742649; fax +39 0584 742592

Cipollino Cremo Tirreno

ESTHETIC CHARACTERISTICS

A continuous run of dark green veins over a pastel grass-green ground is Verde Antigua's main characteristic, creating a successful whole of lively patterns and peaceful colors. The contrast this creates is a new and unusual one, outside the mold, and is returned to the marble realm by its close-knit weave of white veins.

PRODUCTION

Verde Antigua is extracted in a quarry in Iran which produces about 100 to 200 tons per month. Blocks are medium- and large-sized.

WORKMANSHIP AND USES

Verde Antigua's esthetic features show up best when worked in slabs. Surface treatments are polishing or fine-smoothing. Although the material can be used outdoors, it is usually for interiors, for flooring, facing, and bathroom décor. It is often worked in small blocks for bathroom vanity tops. The material has to be reinforced with resin and webbing.

WORKS

Most of the works that have used Verde Antigua are the luxury furnishings of private homes. It was also used to decorate the bathrooms of the prestigious Hotel Villa La Massa in Florence.

Verde Antigua

ESTHETIC CHARACTERISTICS

Ming Green's most outstanding feature is its color, which has no equals in tone and design. The ground looks like a light grass-green mass, fairly bright in places, shading to gray in small, shapeless blots. The whole is pervaded by a weave of green veins that are the same color as the ground but much more intense. The veins may take an irregular course or run parallel to one another, giving the material a definite movement.

PRODUCTION

Ming Green is extracted in China and output is about 600 tons a year. Blocks are small- to medium-sized.

WORKMANSHIP AND USES

Ming Green is mainly worked in slabs or made-to-measure formats for use with differently colored materials to decorate interiors. It is also worked as solid stone for special pieces, and its surface is always polished. The material requires reinforcement with a coat of resin and webbed backing.

WORKS

Ming Green is a fairly recent material and its use, not yet common, has mainly been in luxury buildings. One of the most prestigious is the New Guest Complex at Bait al Barakah in Oman.

Ming Green

ESTHETIC CHARACTERISTICS

An especially elegant marble, Verde Patricia expresses its character in the contrast between the very pale green of the veining and the very dark green of the ground, barely lightened by light green dotting. The veins run straight but without a prevalent direction; often their pattern gives the material a brecciated look and, if very close-knit, it prevails over the ground. Verde Patricia is available in the Classica variety (with a very dark ground) and in Chiara, in which the ground is paler.

PRODUCTION

Verde Patricia is extracted in just one quarry located near Gressoney in the Aosta Valley, Italy. The quarry, worked underground, turns out a yearly average of 500 m^3 in squared blocks. Block size runs from a minimum of 250 x 130 x 80 cm to a maximum of 330 x 150 x 120 cm.

WORKMANSHIP AND USES

Verde Patricia is preferably worked in large slabs, on which its esthetic features are best admired. Surface treatment is always polished. Verde Patricia is especially suitable for flooring and facing and is frequently used in small blocks for special pieces. It is a good idea to give the blocks a seasoning period before sawing.

WORKS

In the 1980s, Verde Patricia was the official material for Cartier boutiques. Among the most significant examples is the one in Place Vendôme in Paris. It was also used to decorate the Casino in San Vincent.

THE COMPANY

Verde Patricia is handled exclusively by Carlo Telara Marmi & Graniti s.r.l. This company, founded in 1925, extracts Bianco Lavagnina, a premium variety of Bianco Carrara, and markets finished and semi-finished pieces of marble and granite in general.

In this latter field Carlo Telara Marmi & Graniti finds its highest expression, as a specialist in supplying nonstandard finished goods for homes. Its main markets for this are the U.S. and northern Europe. Among its most significant commissions are the National Library of Canberra, Australia; some interiors of Harrod's of London; and the ferry terminal in the Port of Genoa, designed by A. Rizzo.

Carlo Telara Marmi & Graniti s.r.l.
Via Carriona 263, 54031 Carrara Avenza (MS), Italy
ph. +39 0585 857351; fax +39 0585 50198
Internet: http://www.telara.com; e-mail: telara@telara.com

Verde Patricia

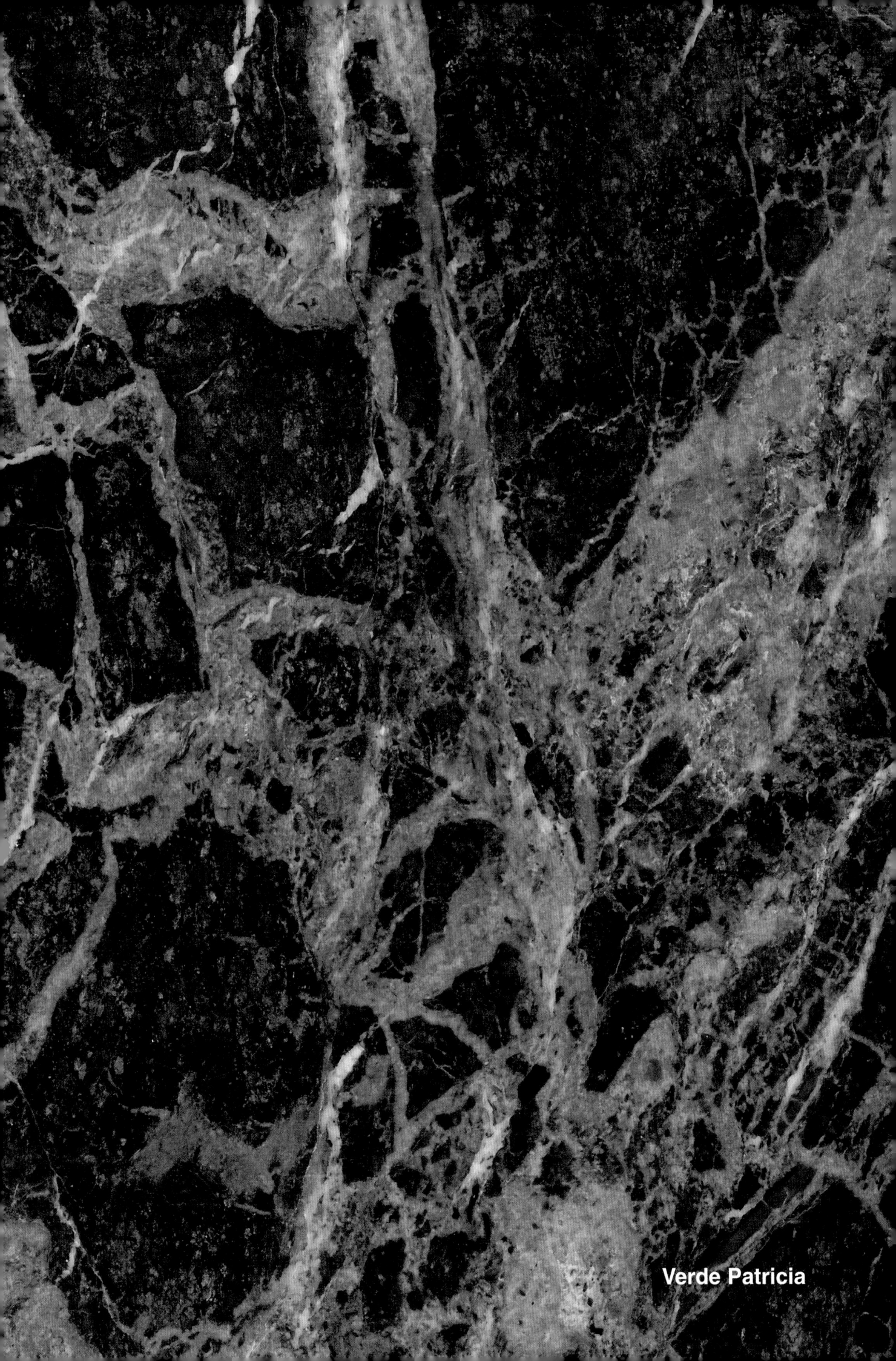

ESTHETIC CHARACTERISTICS

Among green marbles, Verde Antico d'Oriente is primarily distinctive for its fine weave of pale green veining that almost forms a threadlike, sometimes cloudy, web that rises from the dark green ground. Frequently the pattern is directional, giving the material just a hint of movement. Depending on the darkness of the ground, but especially on the frequency of the veining, Verde Antico d'Oriente is differentiated into dark, medium, and light—subvarieties in which the overall green is seen as darker or lighter. These subvarieties are quite constant in their esthetic features.

PRODUCTION

Verde Antico d'Oriente is produced in just one quarry, which guarantees constant supply of about 100 m^3 a month. The blocks are always squared, devoid of structural defects in general, and the average size is between 4 and 5 m^3.

WORKMANSHIP AND USES

Verde Antico d'Oriente is worked in slabs averaging 250 x 130 cm and in tiles. The slabs are preferable if one wishes to highlight the slight movement of the pattern. Small blocks are also frequently worked. Usually the material is so compact it needs no ulterior reinforcement, but sometimes, for the sake of prudence, it is best to coat it with resin. The surface is always polished. Verde Antico d'Oriente is used in luxury interiors for flooring and facing and special pieces (tops, curbing, etc.).

THE COMPANY

Verde Antico d'Oriente is handled exclusively by Block & Rock s.r.l., which markets it in raw blocks and slabs. Founded in 1993, Block & Rock immediately caught the international market's attention for its great organizational abilities and business dynamism. It makes special efforts in searching for new materials (of which Verde Antico d'Oriente is a prime example) in order to offer its clientele a varied and up-to-date assortment of materials that meet with growing market demands.

Block & Rock s.r.l.
Via Antica Massa 38, 54031 Avenza, Carrara (MS), Italy
ph. +39 0585 52883 r.a.; fax +39 0585 856708
Internet: http://www.blockrock.com; e-mail: blockroc@tin.it

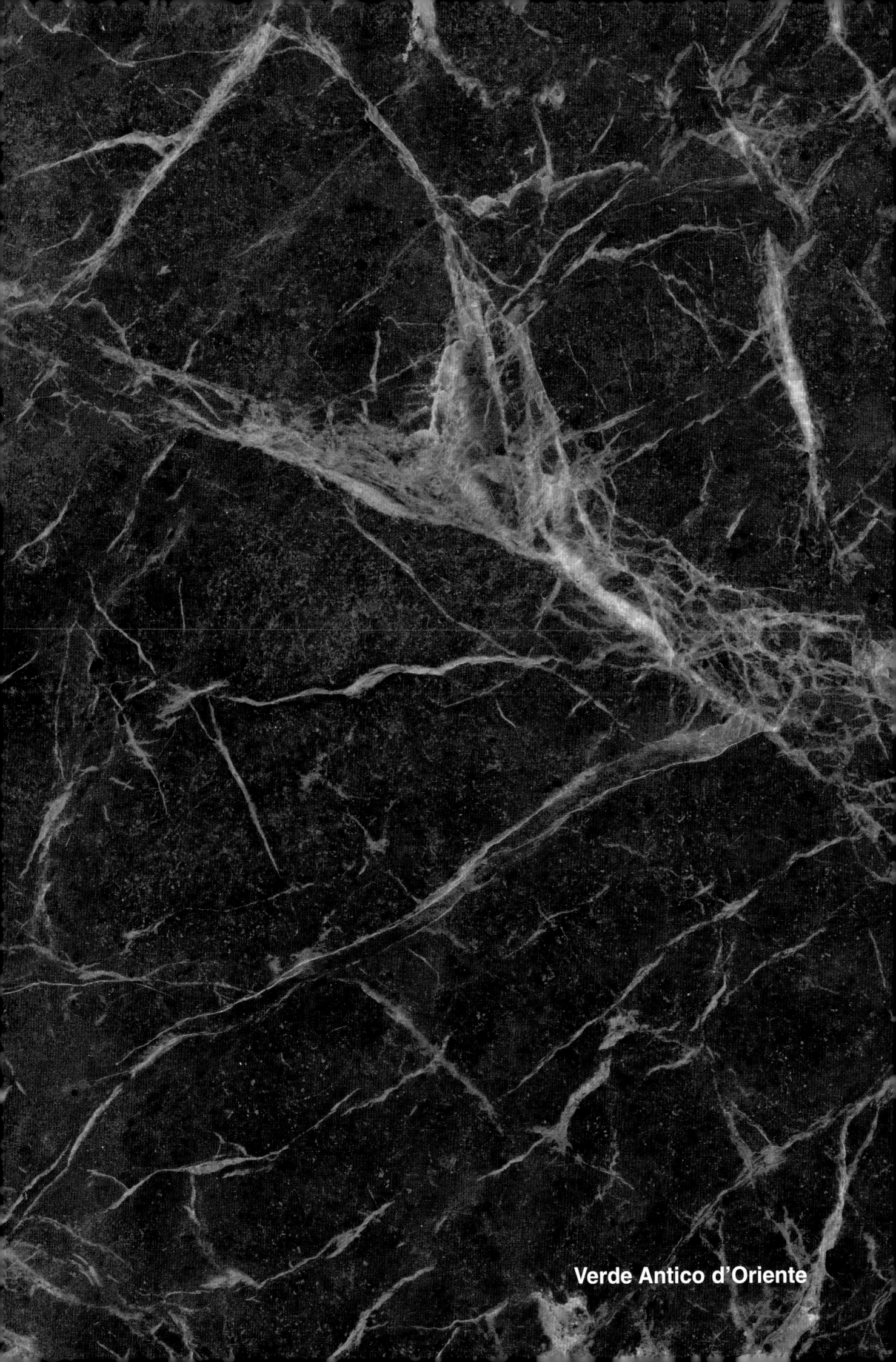

Verde Antico d'Oriente

ESTHETIC CHARACTERISTICS

The special feature of Verde S. Denis is its weave of white or pale green veining, which often forms rounded links with clear-cut or shaded outlines that fringe into the dark green ground. The pattern may become so dense that it thoroughly incorporates the small parts of the ground and often hints at a delicate orientation that greatly enhances the material's esthetic worth.

PRODUCTION

Verde S. Denis is extracted from just one quarry near Gressoney in the Aosta Valley. The quarry has been operating since 1924 and turns out an average of about 200 tons a month in squared, medium- to large-sized blocks. Given its geographical location, the quarry does not operate in winter.

WORKMANSHIP AND USES

Verde S. Denis is worked mainly in large slabs for interiors and as solid stone for special works like tops, curbing, etc. Applications are primarily internal but, if treated, the material can also be used outdoors. The blocks must be seasoned before use and slabs are usually resin-coated and reinforced.

WORKS

Among the most recent and significant works created with Verde S. Denis are the Ranieri Museum in Monte Carlo and the Venetian Hotel in Las Vegas. Verde S. Denis has also been frequently chosen by Tiffany & Co. and Ritz Carlton for their projects worldwide. It is also used in the J. McDonald residences in Palm Beach, Thulman Eastern, and Baltimore.

THE COMPANIES

Verde S. Denis is one of the top materials sold by Italstone Industries s.r.l., the Italian affiliate of Techstone System of Palm Beach, Florida. Italstone Industries specializes in supplying made-to-measure marble and granite for internal coverings and cladding. Among its major commissions are the Magic Kingdom Hotel at EuroDisney in Paris; the Portofino Bay Hotel at Universal Studios in Orlando, Florida; and the Ritz Carlton Hotel in Palm Beach

Italstone Industries
Viale Roma, 20 54100 Massa (MS) Italy
ph. +39 0585 810127; fax: +39 0585 810272
Internet: http://www.italstoneindustries.com; e-mail: italstone@italstoneindustries.com

Valta s.r.l.
Via Covetta, 1 54031 Avenza Carrara (MS) Italy
ph. +39 0585 856369; fax: +39 0585 856371
e-mail: valtasrl@tin.it

Verde S. Denis

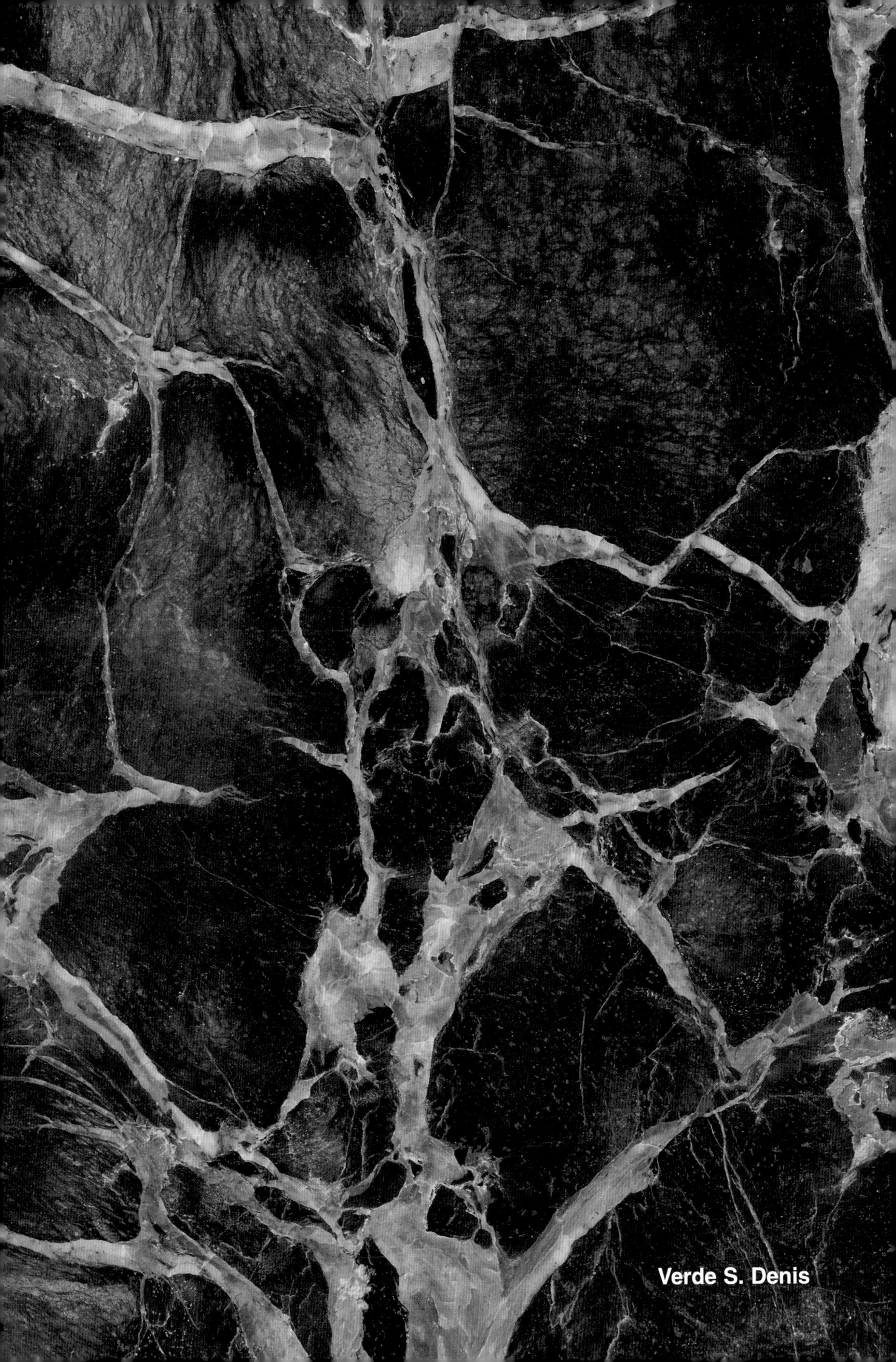

ESTHETIC CHARACTERISTICS

On a very dark green, nearly black, ground, uniform in tone and intensity, runs a thin, sparse weave of veining of a paler, sometimes bright, green that gives the marble its name of Verde Rameggiato (Branched Green). It is exactly this contrast with the veining that gives the material luminosity, while its dark ground makes Verde Rameggiato elegant and delicate at the same time.

PRODUCTION

Verde Rameggiato is extracted in a quarry near Chambave in the Aosta Valley. Annual production is about 1,000 m^3 in medium- to large-squared blocks.

WORKMANSHIP AND USES

Verde Rameggiato is worked both in large-size slabs and in tiles, always polished and destined mainly for flooring and facing. It is quite often worked in small blocks for special jobs. Especially distinctive among these works, with great esthetic effect, are columns covered with large contoured elements. The material needs to be reinforced with resin and sometimes with webbing. A seasoning period is required before working it.

WORKS

Some of the most prestigious works created with Verde Rameggiato are the main offices of the Banca d'Italia at the Centro Donato Menichella in Frascati and the United States Court House in Portland, Oregon.

THE COMPANY

The production of Verde Rameggiato is handled exclusively by Cogemar Marble & Granite s.r.l. Founded in 1980, the company is part of the Cogemar Group which includes three companies based in Massa and several international affiliates. As a whole, the group covers all the fields in the stone industry, from extraction (the group also owns a Bianco Carrara quarry and holds an exclusive for various other materials) to the production and installation of the finished product, even made to measure. Where works are concerned, the Cogemar Group has supplied to dozens of prestigious projects worldwide.

Cogemar Marble & Granite s.r.l.
Via Aurelia Ovest 355/a, 54100 Massa (MS) Italy
ph. +39 0585 8360; fax +39 0585 830341
Internet: http://www.cogemar.com; e-mail: cogemar@cogemar.com

Verde Rameggiato

ESTHETIC CHARACTERISTICS

What is striking about Cavendish Antique is its very dark green ground, which just visibly shows a very slight weave of paler shades. The shadings enlarge in places, becoming isolated speckles or true pale green veining, uneven in shape and thickness and spread over the ground in no evident pattern.

PRODUCTION

Cavendish Antique is extracted in a quarry in Vermont. Production is about 60 m^3 per month in blocks that are always squared and large.

WORKMANSHIP AND USES

Thanks to its esthetic features, Cavendish Antique is normally used in both slabs and tiles. These products are generally resin-coated and reinforced with webbing. The material is frequently used in small blocks for sills, architraves, tops, and other special works. It can also be used outdoors, especially as cladding, without harm to its mechanical properties. It is usually installed in interiors as flooring.

WORKS

Cavendish Antique has been used in the Public Library in Washington, D.C. and the Vermont National Bank in Rutland, Vermont.

THE COMPANY

Cavendish Antique is produced by F.lli Mazzucchelli Marmi s.r.l. In addition to its extraction department (which also includes a Bianco Carrara quarry and shareholdings in a Danby Marble quarry in the U.S.), the company markets marble and granite generally, both raw and as finished products and even supplies products made to measure.

F.lli Mazzucchelli Marmi s.r.l.
Via Ilice 20, 54031 Nazzano Carrara (MS), Italy
ph. +39 0585 51085; fax +39 0585 51228
e-mail: vince.mazzucchelli@libero.it

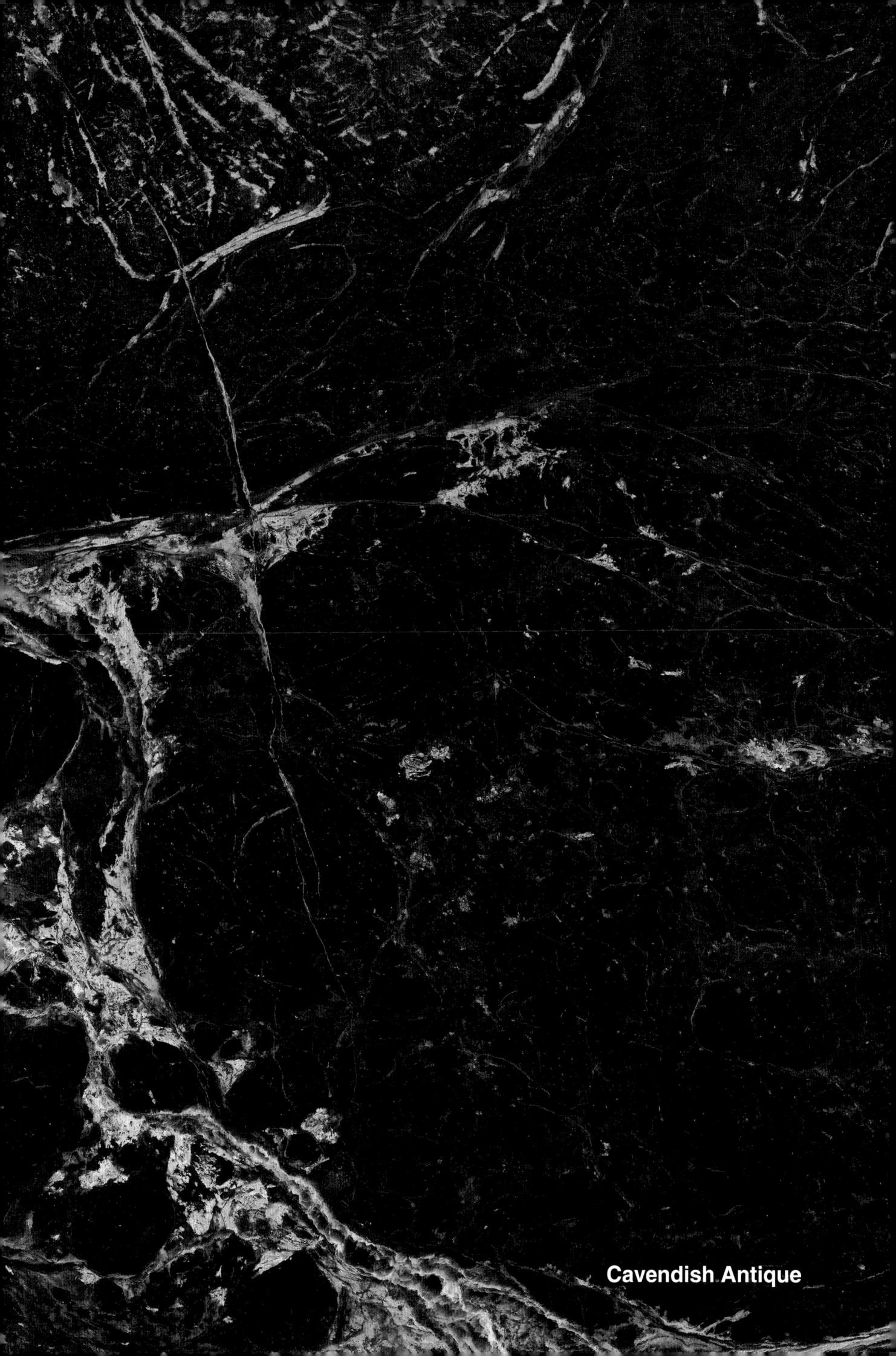

Cavendish Antique

ESTHETIC CHARACTERISTICS

Irish Connemarble Green has a sequence of green bands, extraordinarily changeable in tone and intensity, that run parallel to one another without interruption, sometimes shading into each other and sometimes with clear-cut edges generating strong color contrasts. From the paler bands, veining of the same color branches out into the darker areas, forming an elegant, wide-linked web. The resultant picture is certainly unique of its kind, a well-defined esthetic model yet one that generates patterns that are never identical.

PRODUCTION

Irish Connemarble Green is extracted from a quarry in Ballynahinch in County Galway, Ireland. Annual production is about 10,000 tons, mostly in blocks for sawing and to a lesser extent in shapeless pieces for artisan and jewelry work.

WORKMANSHIP AND USES

Irish Connemarble Green is worked in both slabs and tiles. In slabs the material fully expresses its esthetic worth and is very effective in mirror-image installations. It is recommended for interiors only. The material needs to be resin-coated and reinforced with webbing.

HISTORY

According to a number of sources, Irish Connemarble Green was extracted as early as the end of the first millennium. What is certain is that it was used to decorate many European palaces during the Baroque and Rococo periods.

WORKS

Irish Connemarble Green was used to decorate the Vatican and the palace of Versailles. It was recently used at Trinity College in Ireland and at the University Club in New York City.

THE COMPANY

Irish Connemarble Green is distributed exclusively by Antolini Luigi & C. S.p.A. This company, founded in 1956, specializes in supplying marble, granite, and stone as semi-finished and finished products and, through its subsidiary, Eurotrading S.p.A., as blocks. The company normally offers over 450 varieties of materials, many exclusively, and owns a Rosa Beta quarry and another of Serizzo Antigorio. Its four processing plants have a total of twenty-seven gangsaws and six slab lines combined with many machines for processing and special treatments.

Antolini Luigi & C. S.p.A.
Via Marconi 101, 37010 Sega di Cavaion (VR) Italy
ph. +39 045 6836611; fax +39 045 6836745
Internet: http://www.antolini.it; e-mail: marketing@antolini.it

Irish Connemarble Green

ESTHETIC CHARACTERISTICS

A material with a special look, Breccia Paradiso is composed of a dark brown ground, which is crossed by a close-knit weave of pale, thin hazelnut veins. The veins, which generally run straight, often thicken to form a dense, uneven network that incorporates the ground mass in individual elements, giving the material a brecciated appearance. This creates a picture that is muted in color but lively in form—a singular combination that is a prime feature of this marble.

PRODUCTION

Breccia Paradiso is extracted from a quarry located near Esperia, in the province of Frosinone, Italy. Their output is about 300 m^3 in squared blocks sized 280 to 310 x 140 to 150 x 120 to 150 cm.

WORKMANSHIP AND USES

Breccia Paradiso can be worked as both slabs and tiles without jeopardizing its esthetic characteristics. It is most often used for flooring and cladding and is generally polished. However, it is frequently honed and, when used for outdoor elements, is bushammered.

WORKS

Among the most prestigious works using Breccia Paradiso are the World Trade Center (Twin Towers) in New York City; the Hyatt and Marriott Hotels in New York City; the TWC Teleport Business Park in Bogotá (the largest project built in Colombia); and the Hotel Puerta del Camino in Santiago de Compostela, Spain.

THE COMPANY

Breccia Paradiso is produced exclusively by Unioncave s.r.l. In addition to extraction work, the company also processes blocks into slabs, modulmarmo, cut-to-size elements, and special pieces. The company also sells many varieties of marble and granite.

Unioncave s.r.l.
Via Leopardi 23, 03043 Cassino (FR), Italy
ph. +39 0776 910614; fax +39 0776 910548
e-mail: mvalente@officine.it

Breccia Paradiso

ESTHETIC CHARACTERISTICS

A sky so blue it reaches tones of lapis lazuli, and so clear that only here and there is it covered with tiny white clouds that increase its depth. This is the picture painted by Royal Sapo Sodalite Blue Royal, peerless in the world of ornamental rocks and often enriched with a thin golden weave that stands out from the blue ground. When the white clouds increase to predominate over the blue, the variety is called Sodalite Blue Nuvolata.

PRODUCTION

Sodalite Blue Royal is extracted from a quarry in the Bolivian Andes at an altitude of about 4,000 m. Total output (of the two varieties described) is about 100 tons per month of squared blocks sized between 120 x 90 x 90 and 300 x 160 x 140 cm.

WORKMANSHIP AND USES

Sodalite Blue Royal can be worked as slabs or tiles for facing and flooring. The material generally has to be reinforced with resin and webbing. Currently, dimension work with it is uncommon, but there are good bases for its development. In any case, it is recommended for interiors.

WORKS

Sodalite Blue Royal appeared on international markets only recently and has been used essentially in luxurious private homes.

THE COMPANY

Sodalite Blue Royal is distributed exclusively by International Italmarmi di Nicastro R. & C. s.r.l. Founded in 1981, it derives from another company working in the stone sector since 1976. The company handles and markets the production of slabs and finished pieces in marble and granite. Along with those cited, some of its major commissions were the precast flooring for the Langham Hilton Hotel (the only six-star hotel in England), and the finished material for the Portman Ritz Carlton Hotel in Shanghai. International Italmarmi works with a quality system conforming to UNI EN ISO 9002 standards (1994) as officially certified by Sincert/Det Norske Italia s.r.l.

International Italmarmi di Nicastro R. & C. s.r.l.
Viale Roma 209 bis, 54100 Massa (MS), Italy
ph. +39 0585 254355; fax +39 0585 254307
Internet: http://www.internationalmarmi.com; e-mail: info@internationalmarmi.com

Sodalite Blue Royal

ESTHETIC CHARACTERISTICS

The appearance of Bardiglio Cappella is undoubtedly singular and fascinating, evoking images of laser beams that cut the cosmic darkness. Clearly standing out from a dark gray ground, variable in intensity, are luminous stretches of lines interrupted by tiny, sudden bursts of white light. The gray ground and white lights run straight, with certainty, giving the material not only a well-defined pattern but also a feeling of depth that accentuates its movement.

PRODUCTION

Bardiglio Cappella is extracted in the Versilia marble basin in Tuscany's Apuan Alps. The quarry, located in La Cappella (thus, its name) has inconstant production that reaches an average of 200 tons per month. Blocks are both shapeless and squared, and the latter run in size from 150 x 80 x 100 cm to 250 x 150 x 200 cm. Larger blocks in particular may have fractures, reducing the measurements of the intact slabs that can potentially be produced.

WORKMANSHIP AND USES

Bardiglio Cappella is generally worked as slabs and tiles for flooring and facing. As mentioned, intact slabs are generally sized smaller than the blocks. A typical use of tiles is in checkerboard flooring together with white marble. Surface treatments are usually polished or smoothed. A combination of polishing and sanding greatly accentuates the white/black contrast already found in the material's chromatic range.

HISTORY AND WORKS

The first use of Bardiglio Cappella is uncertain but was undoubtedly early, at least in the nineteenth century if not before (Stoppani mentions its quarrying since the eighteenth century). The material was used in the past for the pavements and decorations of many old churches, while today it is used to recoup and restore them.

THE COMPANY

Bardiglio Cappella is handled exclusively by SO.LU.BER s.r.l. Founded in 1983, the company is not only the main supplier of raw Calacata Luccicoso, but also produces slabs of white and colored marbles and stones in general and finished products. One of its main commissions was supplying raw and finished Calacata Luccicoso for the Al Hani Tower in the United Arab Emirates.

SO.LU.BER. s.r.l.
Viale Puccini 30, 54100 Massa (MS) Italy
ph. +39 0585 489487; fax +39 0585 45760

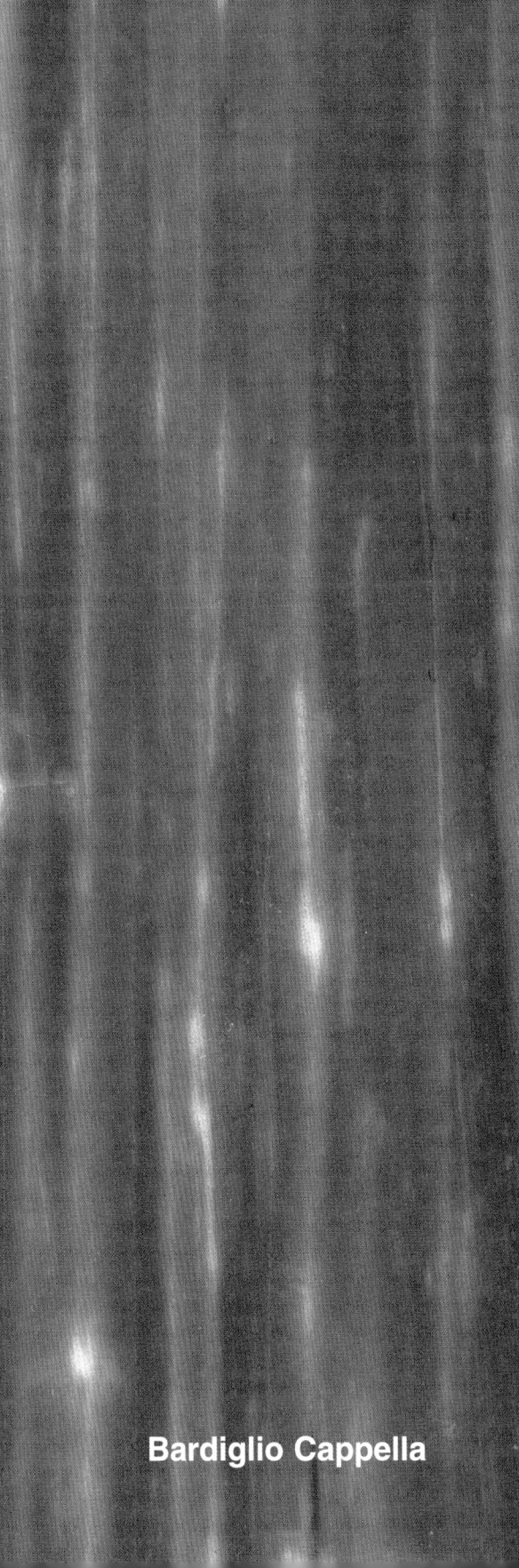

Bardiglio Cappella

ESTHETIC CHARACTERISTICS

Uniform color is doubtless one of Pietra di Matraia's features, but making it unique is its particular tenacity, highly unusual for a sandstone destined for ornamental purposes. A homogeneous, medium-fine grain sandstone, which, when freshly cut looks dark iron gray. Very rare and of small size are smoke-black, fine-grain spots.

PRODUCTION

Pietra di Matraia is extracted in several quarries on the uphill slope of the homonymous town in the province of Lucca, Italy. Output is inconstant, since work is solely by commission. Production capacity is about 400 tons a month in blocks 350 x 170 x 170 cm.

WORKMANSHIP AND USES

Pietra di Matraia is typically worked in small blocks. The articles most commonly made with it are thresholds, sills, landings, cornices, stairs, and fireplaces. In regard to special works, the material's high resistance enables it to be carefully carved into fine details such as a small relief or thin border, without compromising the integrity of the work. It is also frequently used as curb- and paving-stones with a ribbed surface, but production includes slabs for flooring and paving as well. In the first case the surface finish is usually fine-smoothed, although the material can be perfectly polished. For outdoor use, excellent results are achieved with various shock treatments, and even with flaming, which generates a rough surface while maintaining the original color.

HISTORY

Pietra di Matraia has been used since the Renaissance. It seems it was first excavated in the mid-sixteenth century to build the Matraia bell-tower, still in excellent condition. Immediately after that time it began to be used in nearby Lucca both for urban fixtures, paving in particular, and for private buildings.

WORKS

Pietra di Matraia has been used mainly in recouping and restoring buildings and historic works such as the paving in Piazza dei Miracoli and the Torre Guelfa in Pisa, Piazza Grande in Lucca, Piazza Civini in Pistoia, the Pontestrada bell-tower, and the City Hall in Frassinoro. Important modern works include two banks, one in Singapore and one in Dubai.

THE COMPANY

Pietra di Matraia is extracted and processed by Matraia s.r.l. Although founded only recently, this company has the skills and professionalism that only longstanding stoneworking traditions can provide. In fact, the company is simply the latest expression of a structure devoted to sandstone production and processing since 1400, whose works have been enriching Lucca's architectural heritage for centuries. Aiming at maintaining the quality and creativity typical of artisan work, Matraia s.r.l. works solely by commission.

Matraia s.r.l.
Via Traversa 133, 55013 Lammari (LU), Italy
ph. +39 0583 436066 / 402217; fax +39 0583 402215

Pietra di Matraia

ESTHETIC CHARACTERISTICS

The prime feature of Nero Piemonte di Ormea is its deep black ground, which gives the material particular elegance and worth. At times the depth of the background color, which is the result of its chromatic purity as well as its very fine grain. This marble is greatly accentuated by a delicate weave of grayish shadings that may or may not turn into short white, clear-cut veins.

PRODUCTION

Nero Piemonte di Ormea is extracted in Piedmont, Italy, in the province of Cuneo. The quarry lies on the southwestern slope of the mountains in which the Tanaro River originates. Total production is 8,000 tons per year, 35 percent of prime quality. 50 percent of output is in squared blocks which average 250 x 140 x 140 cm in size; 25 percent is in semi-squared blocks (220 x 120 x 120 cm) and the remaining 25 percent in shapeless blocks weighing an average of 10 tons.

WORKMANSHIP AND USES

The worked product generally consists of small to medium-sized polished slabs for flooring and facing, in which the great elegance of the material can be fully admired. Also frequently produced are flooring tiles, which can be polished, smoothed, chiseled, or sanded. In cases of internal and external historical restoration, preference is usually given to an antiqued surface. And here the product can be used as paving stones, cubes, and large slabs. The material may need to be reinforced and resin-coated.

WORKS

Nero Piemonte di Ormea was used in the past for a great many works that have become part of the historical heritage of Piedmont and Liguria. More recently, it was used in restoring the pavement of the church of S. Antonio Abate di Priero (Cuneo); for paving the historical centers of Genoa, Borgomaro (Imperia), Finalborgo, and Noli (Savona); on the Bossolasco (Cuneo) City Hall; for the bridge in the historical center of Carcare (Savona); and for the Corni and San Pietro bridges, both in Ormea.

THE COMPANY

Nero Piemonte di Ormea is extracted and processed exclusively by Zoppi s.r.l. in Priero in the province of Cuneo. Working since 1968 in building materials, in the 1990s the company decided to reactivate the Nero Piemonte di Ormea quarry, which was in disuse for years due to an unfavorable local economic situation. Presently, Zoppi s.r.l. goes through the entire processing cycle in its own plants in Priero, from block sawing to finished products of various types. The works mentioned above were all done by Zoppi srl which, with its high-quality materials and high-speed production, is an important reference point in the international supply of quality black marble.

Zoppi s.r.l.
Via Roma, 2 Priero (CN) Italy
ph. +39 0174 79107; fax +39 0174 79931
e-mail: Zoppi.srl@isiline.it

Nero Piemonte di Ormea

ESTHETIC CHARACTERISTICS

Among black marbles Noir Saint Laurent is distinctive for its warm and elegant color and the subtle weave of its design. On a black ground (at times, having slightly brownish shadings) run delicate white or golden veins that are always rather fine and generally straight, although without a preferential direction.

PRODUCTION

Noir Saint Laurent is extracted from a quarry in the south of France. Monthly production is about 150 tons, 50 percent of which are top quality. Squared blocks, which account for 30 percent of output, are sized a maximum 280 x 150 x 80 cm.

WORKMANSHIP AND USES

The best format for valorizing the esthetic qualities of Noir Saint Laurent is polished slabs for use as flooring and facing. It is also frequently made to measure and used with materials of another color. A smoothed surface is possible but less common. The worked product has to be coated with resin, stuccoed, and reinforced with webbing.

WORKS

Noir Saint Laurent has been used in many important works. Among them are the recently built Kowloon Station in Hong Kong, the Conrad Hilton Hotel in Hong Kong, and the Hilton Intercontinental in Athens.

THE COMPANY

Noir Saint Laurent is marketed by Byblos Stone s.r.l. of Marina di Carrara (Massa), which is its exclusive dealer for Italy. In addition to Noir Saint Laurent, the company (founded in 1970) also supplies as raw or finished products a wide range of premium materials extracted in France, the main varieties of Apuan marbles and Italian and foreign granites. Among its major commissions was the aforementioned Kowloon Station in Hong Kong.

Byblos Stone s.r.l.
Via W. Muttini 22, 54036 Marina di Carrara (MS) Italy
ph. +39 0585 786515; fax +39 0585 631127
e-mail: byblostone@tin.it

Noir Saint Laurent

ESTHETIC CHARACTERISTICS

Portoro is characterized by delicate veins running over a deep black, fine-grained ground, threadlike in some places and in large roundish spots in others, but in any case with a slight and graceful undulation. The veining is often golden yellow (from whence the name Portoro came) or, infrequently, white. The veins run parallel to one another, giving the design a definite orientation.

PRODUCTION

Portoro is extracted from a few quarries near Portovenere in the province of La Spezia. Total annual production is just over 1,000 tons, in medium- and large-sized squared and shapeless blocks.

WORKMANSHIP AND USES

Portoro is often used as cut-to-size product in combination with other materials. It is also frequently worked in small blocks for special pieces and in medium-sized slabs for flooring and facing. Its use should be limited to interiors. The material usually has to be coated with resin and reinforced with webbing.

HISTORY

In all likelihood, the extraction of Portoro began in the Renaissance when Cosimo de Medici I spurred a search for colored marbles. It was exploited much more in the past than now and was sold in a number of subvarieties that today have mostly disappeared.

WORKS

Portoro is quite popular worldwide. It was used to decorate the Melbourne Casino in Australia, and in Italy for the interiors of the Hotel Principe di Savoia in Milan designed by Maurizio Papiri.

Portoro

ESTHETIC CHARACTERISTICS

Black Pearl, also known as Abu Black, has unique esthetic features. To the typical black static materials or those slightly moved by vein direction, it adds a markedly moving pattern developed by its gray and white veins. These stand out from the black ground, often in the guise of narrow but continuous bands of undulating color that frequently fold into themselves. When cut with the grain, the veins get wider, giving the material a fascinating cloudy pattern.

PRODUCTION

Black Pearl is extracted in the state of Rajasthan in India. The quarries, located near the town of Abu, turn out up to 150 m^3 of squared blocks a month. The blocks have an average size of 200 x 100 x 100 cm.

WORKMANSHIP AND USES

To valorize Black Pearl's appearance to the utmost it is best worked in slabs. In this case fine results can be achieved with mirror-image installations. The material, which mantains polishing also in exteriors, is very popular as tiles and for special solid pieces such as vanity tops and columns. Its surface is usually polished or antiqued. It may need reinforcement with polyester resin.

THE COMPANY

Black Pearl has been produced and marketed by Trivedi Crafts Pvt Ltd. since 1937. The company was created to exploit the Ambaji quarries in order to rebuild the area's ancient temples and is now a leading part of the Trivedi Marble Group. The group's work runs from extracting to supplying finished products, including special works. It also owns an Ambaji White quarry and another of Rajasthan Green. It has a tile plant in Ambaji, a slab plant in Ahmedabad and another in Ahmedabad for special works (carved products). Thanks to its up-to-date architecture and design division, the company can realize projects for any type of commission. One of its most important was the Swaminarayan Temple in Neasdon, near London.

Trivedi Crafts Pvt Ltd.
68 Premanjali Society, Bodakdev, Ahmedabad 380054 (India)
ph. +91 79 674 5522/3110/4510; fax +91 79 676 6220;
Internet: http://www.trivedi-marble.com; e-mail: trivedi_kiran@vsnl.com

Black Pearl

San Miniato al Monte, Florence

PART III

PORTFOLIO OF BUILT WORKS

Ambaji White

Delwara Temple

Mt. Abu Rajastan (India) circa 800
Renovation: 1950
Arch.: Shri Amratlal Trivedi

Delwara Temple

Mt. Abu Rajastan (India) circa 800
Renovation: 1950
Arch.: Shri Amratlal Trivedi

Bianco Acquabianca

Kiarong Mosque

Sultan of Brunei
Arch.: Zaini
Supplier Cia-Mar (Carrara)

Imperial Danby

Supreme Court

Washington, D.C. (USA)
Arch.: Cass Gilbert

Imperial Danby

Thomas Jefferson Memorial

Washington, D.C. (USA)
Arch.: John Russell Pope

Statuario Michelangelo

Altar of Presbytery of Padua Cathedral

Padua (Italy)
Sculptor: Giuliano Vangi
Supplier: Cave Michelangelo (Carrara)

Arabescato Piana

Private villa

Arles (France)
Supplier: Mirko Menconi Marmi (Carrara)

China Telecom

Shanton (China)
Supplier: Mirko Menconi Marmi (Carrara)

Arabescato Piana

Private villa

Varadero (Cuba)
Supplier: Mirko Menconi Marmi (Carrara)

Breccia Capraia

Home Savings of America Tower

Los Angeles, California (USA)
Arch.: AC Martin Partners
Supplier: Adolfo Forti (Carrara)

Breccia Capraia

Home Savings of America Tower

Los Angeles, California (USA)
Arch.: AC Martin Partners
Supplier: Adolfo Forti (Carrara)

Calacata Borghini

Cassa di Risparmio di Carrara

Carrara (Italy)
Supplier: Imarmi (Carrara)

Trinity Tower Poject

London (U.K.)
Supplier: Cogemar (Massa)

Calacata Borghini

Cassa di Risparmio di Carrara

Carrara (Italy)
Supplier: Imarmi (Carrara)

Calacata Luccicoso

One Market Plaza

San Francisco, California (USA)
Arch.: Cesar Pelli & Associates
Supplier: Carli Cav. Oreste & C. (Carrara)

Statuario

Sillogismo

Arch.: Aldo Rossi
Supplier: Up&Up (Massa)

Calacata Sponda

Brent Cross Commercial Centre

London (U.K.)
Supplier: International Italmarmi (Massa)

Calacata Sponda

Brent Cross Commercial Centre

London (U.K.)
Supplier: International Italmarmi (Massa)

Azul Imperial

Volksbank

Nurnberg (Germany)
Supplier: Rossittis (Holzwickede)

Azul Imperial

Volksbank

Quarzite Blu

Bargetti + Biberstein Office

Solothurn (Switzerland)
Supplier: Solmar (Como)

Quarzite Blu

Private villa

Lecco (Italy)
Supplier: Solmar (Como)

Azul Macaubas

Head office of Ambient Inc.
....

Tokyo (Japan)
Arch.: Aldo Rossi and Cappa Kitai Arch.
Supplier: Cogemar (Massa)

Rosso Carpazi

Quartier 108
....

Berlin (Germany)
Arch.: Van den Valentyn - Matthias Dittmann
Supplier: UpGroup (Massa)

Rosso Verona

Four Millbank

London (U.K.)
Supplier: Cogemar (Massa)

Breccia Pernice

Piccadilly Centre

Sidney (Australia)
Supplier: EssegiMarmi (Verona)

Breccia Pernice

Trump Tower
....
New York City (USA)
Arch.: Swanke Hayden Connell Architects
Supplier: EssegiMarmi (Verona)

Breccia Pernice

Hotel Principe di Savoia

Milano (Italy)
Arch.: Maurizio Papiri
Supplier: Cogemar (Massa)

Pietra Dorata

Malpensa 2000 airport

Milan (Italy)
Supplier: Solmar (Como)

Pietra Dorata

Nymius

Milan (Italy)
Supplier: Solmar (Como)

Giallo Siena

Hotel Principe di Savoia

Milano (Italy)
Arch.: Maurizio Papiri
Supplier: Cogemar (Massa)

Giallo Siena

Royal Inn Holiday

Singapore
Supplier: Cogemar (Massa)

Giallo Siena

V.I.P. Club
....

Singapore
Arch.: Steven J. Leach
Supplier: Elle Marmi (Carrara)

Verde Patricia

Private villa

Carrara (Italy)
Supplier: Carlo Telara Marmi e Graniti
(Carrara)

Verde Rameggiato

SOCIETY IS JUSTICE

Juror Assembly Room

United States Court House

Portland, Oregon (USA)
Supplier: Cogemar (Massa)

Rosso Verona

Hotel Principe di Savoia

Milan (Italy)
Arch: Maurizio Papiri
Supplier: Cogemar (Massa)

Verde S. Denis

**Created by
Giuseppe Garibaldi**
....

Stone Artisan
Pietrasanta (Italy)

Nero Piemonte di Ormea

Chiesa di S. Antonio Abate

Abate di Priero, Cuneo (Italy)
Supplier: Zoppi (Cuneo)

Nero Piemonte di Ormea

Centro storico

Noli Sarona (Italy)
Supplier: Zoppi (Cuneo)

Pietra di Matraia

Piazza Civinini

Pistoia (Italy)
Supplier: Matraia (Lucca)

Torre Guelfa

Pisa (Italy)
Supplier: Matraia (Lucca)

Pietra di Matraia

Piazza dei Miracoli

Pisa (Italy)
Supplier: Matraia (Lucca)

Portoro

Hotel Principe di Savoia

Milan (Italy)
Arch: Maurizio Papiri
Supplier: Cogemar (Massa)

Noir Saint Laurent

The Essex House Hotel

New York City (USA)
Arch: Pierre Yves Rochon
Supplier: Innovative Marble & Granite (NY)

BIBLIOGRAPHY

AA. VV. *Marmi Antichi*. Rome: Edizioni de Luca, 1997.

G. Blanco. *Dizionario dell'Architettura di Pietra. I Materiali*. Rome: Carocci Editore S.p.A., 1999.

———. *Pavimenti e Rivestimenti Lapidei*. Rome: La Nuova Italia Scientifica, 1994.

———. *Pavimenti in Pietra*. Rome: La Nuova Italia Scientifica, 1994.

F. Bradley. *InfoMARBLE*. Milan: Promorama, 2000.

———. *La Scelta delle Rocce Ornamentali in Architettura, Caratteristiche Estetiche Tecniche e Commerciali/ The Choice of Dimension Stone in Architecture, Aesthetic, Technical, and Commercial Characteristics*. Florence: Studio Marmo, 1997.

———. *Natural Stone: A Guide to Selection*. New York: Norton, 1998.

M. Giornetti. *Glossario tecnico del settore lapideo*. Carrara: Internazionale Marmi e Macchine, 1991.

Marble Institute of America. *Dimensional Stone. Volume I*. Farmington, MI: Marble Institute of America.

———. *Dimensional Stone. Volume II*. Farmington, MI: Marble Institute of America.

M. Pieri. *Marmologia. Dizionario di Marmi e Graniti italiani e esteri*. Milan, Hoepli, 1966.

———. *Pigmentazione e tonalità cromatica nei marmi*. Milan: Hoepli, 1957.

E. M. Winkler. *Stone in Architecture*. Berlin: Springer-Verlag, 1994.

INDEX OF FINE MARBLES

Ambaji White   50
Arabescato Piana   58
Azul Imperial   78
Azul Macaubas   76
Azzurro d'Oriente   74
Bardiglio Cappella   134
Bianco Acquabianca   52
Black Pearl   144
Breccia Capraia   60
Breccia Paradiso   130
Breccia Pernice   90
Brèche de Vendôme   94
Calacata Borghini   62
Calacata Luccicoso   66
Calacata Sponda   68
Cavendish Antique   126
Cipollino Apuano   110
Cipollino Cremo Tirreno   112
Diaspro Tenero di Sicilia   92
Fior di Pesco Classico Apuano   98
Giallo Reale   102
Giallo Siena   108
Imperial Danby   54
Irish Connemarble Green   128
Ming Green   116
Nero Piemonte de Ormea   138
Noir Saint Laurent   140
Paonazzo   64
Pietra di Matraia   136
Pietra Dorata   104
Portoro   142
Quarzite Blu   80
Rosa Aurora   72
Rosso Carpazi   82
Rosso Collemandina   86
Rosso Verona   88
Rouge France Incarnat   84
Salomè   100
Sodalite Blue Royal   132
Statuario   70
Statuario Michelangelo   56
Travertino Dorato   106
Travertino Rosso   96
Verde Antico d'Oriente   120
Verde Antigua   114
Verde Patricia   118
Verde Rameggiato   124
Verde S. Denis   122

ABOUT THE CD-ROM

The accompanying CD-ROM contains screen resolution TIFF files for all of the samples in the book. With the appropriate graphics software, the CD images can be used by designers in developing concepts, preparing presentations for clients, and communicating visual information to others. Although the images are primarily intended for on-screen display, they can also be printed on either a black and white or color printer.

Further information about the image formats can be found on the readme.txt file on the CD.

Original images can be obtained from Studio Marmo, 6663 Sedgwick Place, Brooklyn, NY 11220. E-mail: studiomarmo@firenze.net